The Material Culture Reader

The Material Culture Reader

Edited by
Victor Buchli

BERG

Oxford • New York

First published in 2002 by
Berg
Editorial offices:
150 Cowley Road, Oxford, OX4 1JJ, UK
838 Broadway, Third Floor, New York, NY 10003-4812, USA

Berg is an imprint of Oxford International Publishers Ltd.

Library of Congress Cataloging-in-Publication Data
The material culture reader / edited by Victor Buchli.
     p. cm.
Includes bibliographical references and index.
     ISBN 1-85973-554-1 – ISBN 1-85973-559-2 (pbk.)
  1. Material culture. I. Buchli, Victor.
     GN406 .M3493 2002
     306—dc21

                                                        2002012481

British Library Cataloguing-in-Publication Data
A catalogue record for this book is available from the British Library.

ISBN 978 1 85973 554 1 (Cloth)
       978 1 85973 559 6 (Paper)

Typeset by JS Typesetting Ltd, Wellingborough, Northants.

Printed and bound by CPI Group (UK) Ltd, Croydon, CR0 4YY

Contents

Contents

Acknowledgements

I am very grateful to a number of colleagues for their kind help and encourage-
ment while putting together this volume. In particular I want to thank the members
of the Material Culture Group at University College London: Barbara Bender,
Susanne Küchler, Daniel Miller, Christopher Pinney, Michael Rowlands, Nicholas
J. Saunders and Christopher Tilley. Elsewhere I am extremely indebted to Gavin
Lucas, Catherine Alexander, Sarah Jain, Diana Young and Alison Clarke for giving
generously of their time and their invaluable suggestions. Also thanks go to Philip
Wagenfeld for directing me to the works of Richard Caldicott and of course to
Richard Caldicott himself for kindly allowing the use of his haunting images of
Tupperware for the cover. Lawrence Osborne and Ian Critchley at Berg must be
thanked for their steadfast professionalism in the production of the volume but
most of all thanks go to Kathryn Earle for the original idea and her infinite
patience.

List of Figures

Introduction
Victor Buchli

This reader is a compilation of some of the representative works of the Material Culture Group at University College London. It is by no means exhaustive and representative, but it does provide an idea of the range of subjects, contexts and problems material culture studies at University College have addressed over the years and at present. The works here are a sampling of some of the dominant concerns of the contributors. In turn, each contribution is preceded with an introduction by the author placing the work within broader themes relevant to the study of the material world. As a result the compass of these works is quite diverse, giving the reader a sense of the broad and at times conflicting issues in which material cultures studies as a whole participates. What might appear an unruly collection of works is united by an abiding concern for the materiality of cultural life and its diverse and at times conflicting vitality.

Up to now there has never been a 'snapshot' of the work of this group, so the introduction to such a compilation offers a place to look back and try to place this 'snapshot' within the larger scheme of things. As such, this provides the opportunity to examine in general the trajectory of development of material culture studies through a particular cohort of scholars. It also affords the opportunity to attempt and delineate some of the overall issues affecting material cultures studies from this writer's perspective and from there, hopefully, offer some suggestions as to where we are now and where we might be going next.

The particular cohort of which we are speaking are the contributors: Barbara Bender, Victor Buchli, Susanne Küchler, Daniel Miller, Christopher Pinney, Michael Rowlands, Nicholas J. Saunders and Christopher Tilley. It is very obvious that this cohort represents a viewpoint that is distinctly British despite Buchli and Küchler being from the United States and Germany respectively (though they both received their doctorates from British universities). In terms of the British academic traditions of which this group is a part, the cohort is quite firmly situated within the Universities of London and Cambridge and their schools, departments and institutes of archaeology and anthropology. This immediately distinguishes this cohort from scholars of material culture in the United States who, in addition to coming out of the traditions of archaeology and anthropology, are strongly influenced by the tradition of American folklore studies. Bender and Rowlands received their doctorates in archaeology from the Institute of Archaeology, and Saunders from Southampton. All are closely associated with the Institute's current director, Peter Ucko and the legacy of its former director, the Marxist archaeologist Gordon Childe. Similarly, Buchli and Tilley received their doctorates in archaeology from Cambridge having both been supervised by Ian Hodder, who was originally a student of David Clarke's. Miller received his doctorate in oriental studies from Cambridge, but he is closely associated with the group of post-processualist archaeologists who gathered around Ian Hodder. This archaeological leaning within the group is complemented by Küchler and Pinney who both received their doctorates in anthropology from the London School of Economics under the supervision of the anthropologist of art Alfred Gell, a student of Anthony Forge of the London School of Economics.

Material Culture and the Work of Culture

We first encounter the use of the term 'material culture' in English in the nineteenth century. The origins are murky, the first reference to such a concept according to the Oxford English Dictionary was made in 1843 by Prescott on in reference to the 'material civilization' of Mexico in his travelogue. The intellectual history of this concept regrettably is beyond the scope of this introduction, except to say that the study of material culture itself became one of the cornerstones of the nascent independent discipline of anthropology (for a history of the role of material culture in the growth of anthropology see Steadman 1979 and Lowie 1937). In fact, in the late 1800s the concept and its study was almost entirely inseparable from anthropology itself: the so-called 'object-lessons' described by Edward Tylor in his foreword to Ratzel's monumental treatise on the ethnographic study of artifacts, *The History of Mankind* (Ratzel 1896).

However, nineteenth century Victorians who coined the term 'material culture' were by no means the only people preoccupied with artefacts per se. People have always been under their thrall, from palaeolithic assemblages, which seem to suggest an early propensity for collecting, to Babylonian temple collections, ancient Chinese and Roman antiquarians and the cabinets of curios established by Europeans during the Renaissance (Schnapp 1996). In the European context it was these cabinets of curios which were the ancestors of our museums and our pre-occupation with objects in themselves. The history of such collections have been dealt with elsewhere (Belk 2001, Pearce 1994, Thomas 1997). For our purposes here, it is necessary to note that the great Euro-American museums were the institutions in which material culture studies as we know it originally found their home and thrived.

So what has happened in terms of the changing fortunes of material culture studies since the mid nineteenth century? Why was this super-category of objects needed in the first place and why has it fallen in and out of use within anthropology? From its beginnings, material culture as a category and as a field of study was intimately related to larger cultural projects. In the nineteenth century it was used as a way of gauging the degree of technical and social sophistication of a given group. Within these schemes of unilineal evolution European Victorian society was on the top of the scale as the most modern and progressive while other non-European societies descended downwards with various hunter-gatherer groups at the bottom of the scale of human social and technical evolution. This naturally justified European dominance in expansionist imperial affairs, but also served liberal notions of Enlightenment thought which advocated the universality of human experience and justice. The various 'uncivilized' peoples of the world were all subject to the same technical and social processes albeit at different levels, thereby ensuring European imperial dominance. All of humanity's inventions and institutions could be used as an indicator of this inexorable dynamic of inclusive progress.

The emergence of material culture studies was an innovation arising from earlier Enlightenment era preoccupation with the materiality of social life. As Michel Foucault argued in *Space, Knowledge and Power*, the interest in the various material components of social life (i.e. architecture as an aspect of governance) is an eighteenth century preoccupation where 'One begins to see a form of political literature, that addresses, what the order of a society should be, what a city should be, given the requirements of the maintenance of order . . .' (Rabinow 1984:239). The ethnographic urge to order, manage and constitute new political subjects (typically colonial and subject to the principles of universality to which all could aspire to), maintained unilinealism as just such a demonstration of this universal progression. Statecraft, the formation of nationhood and empire were inextricably bound to these quasi-archaeological and ethnographical impulses (Schnapp 1996). These 'objects' of knowledge were vital for establishing the building blocks of statecraft. In short the super-category of objects: material culture, has had from the beginning a utility with specific cultural work to do. As Edward Tylor observed in thinking about the future of material culture studies on the eve of the twentieth century: 'In the next century, to judge from its advance in the present, it will have largely attained to the realm of positive law, and its full use will then be acknowledged not only as interpreting the past history of mankind, but as even laying down the first stages of curves of movements which will describe and affect the courses of future opinions and institutions' (Tylor in Ratzel 1896: xi) .

This emerging understanding of human progress was best expressed in the American anthropologist Lewis Henry Morgan's monumental work *Ancient Society* which laid out the stages of human social and technical evolution from savagery to civilization. Each stage was characterized by a particular level of social and technical achievement which incorporated all peoples and races in the trajectory of human progress. All peoples were alike and uniform and would respond in similar ways given the same technical limitations. Morgan was acutely aware that old life ways were passing through his close work with Native Americans. He could directly witness how older indigenous technical and social achievements were succumbing to the relentless march of Euro-American expansion and progress. Existing peoples in isolation, resembling the earlier stages of humans social evolution existed in Morgan's schema as 'monuments of the past' that is as a living archaeology of early forms of human life. As Karl Marx (a keen reader of Morgan) stated: 'Relics of bygone instruments of labour possess the same importance for the investigation of extinct economic forms of society, as do fossil bones for the determination of extinct species of animals' (Marx 1986: 78). The level of a society was intimately linked to its level of material culture. Thus social organization, social progress could be 'read' from the material culture of a particular people or nation as a fossil could be read to determine stages of the evolution of life on earth. Non-European peoples encountered could be understood within this schema and the differences between nations (especially European nations) could be understood in terms of the differences in their material culture. Thus objects were intimately connected with notions of progress – historically, technically and socially – in short, material culture as it was conceived in the nineteenth century was the modernist super-artefact and the supreme signifier of universal progress and modernity.

Earlier European collections of objects sought to gather the curiosities of the world, both natural and manmade, in cabinets of curiosities. These were not systematized in any particular way, except that they represented everything that was out of the ordinary, exceptional, that is not conventional, be it a tool from a distant society, an unusual rock formation, or natural deformity. The early ethnographic collections that formed the basis of material culture studies in the nineteenth century often contained souvenirs accumulated by sailors on expeditions that one way or another made their way back to the European capitals, forming parts of cabinets of curios or collections dedicated to artefacts of far-flung peoples (Thomas 1991). Beyond mere curiosity, these artefacts and their collections served as proof of an event and contact and knowledge of the peoples encountered. Artefact collections essentially were objectifications of authoritative knowledge (Thomas 1991: 141–3) – and more rarely along with the importation of actually indigenous peoples – no other could possibly do in light of the intellectual and technical circumstances of the time. Thus, as these forms of objectified authoritative knowledge became increasingly unsatisfactory, the random collections of curiosities were superseded by the more systematic collections of later scholars, Haddon, Pitt-Rivers, Boas, etc., These collections were to be rejected again during the careers of pivotal figures such as Boas for not being sufficient objectifications of authoritative knowledge. These objectifications were then supplanted by the ethnographic monograph as it began to emerge through the development of British social anthropology as a source of authoritative knowledge about other societies (Thomas 1991: 141–3). Earlier objectifications of authoritative knowledge were simply superseded by more satisfying techniques – more satisfying in terms of its being able, as Pomian suggests, to render the invisible visible, which he describes as the primary impulse of collecting (Pomian 1990). If initial collections were an attempt to bring such exotic, invisible and otherwise unknowable realms into being for Europeans, then these attempts at knowledge of other realms of experience found more satisfactory expression in the ethnographic monograph which was based on direct field work and participant observation – the souvenir club would no longer suit as an indicator of authoritative knowledge of another realm of experience. As such this requirement has never really been exhausted within anthropology as every Ph.D. student who undergoes the *rite de passage* of fieldwork knows so well.

I do not wish to go over the critical ground covered by others who have emphasized the indisputable ills that have been a consequence of unilinealism and the role of material culture studies within it, except to say that the constitution of mere objects into systems of 'material culture' represented a particular intellectual and political project that required a new kind of conceptual tool: the super-category of objects 'material culture' itself. This project proceeded to materialize precisely such a super-category of objects that never existed before and which was meaningless to the individuals who actually produced these objects. Cook Islanders were hardly producing 'material culture' for the consumption of sailors, travelers, administrators and scholars (as we know from Thomas, Pacific Islanders had very different purposes in mind; see Thomas 1991: 131). Similarly, archaeological excavations constituted a category of objects as 'material culture' entirely foreign to the past producers of these objects. To insist otherwise and claim its ahistorical universality, as many still do, is the act of 'retrofitting' (using Bruno Latour's language) that naturalizes a particular 'concresence' of institutionalized and historically contingent knowledge, which results in his felicitous neologism a 'factish': 'a sustained mode of existence for facts' within a specific 'spatiotemporal envelope' (Latour 1999). This super-category materializes something entirely new and uniquely Victorian and Western, as modern as the artefacts of industrialism on display at the Great Exposition of 1851 from which our more systematic nineteenth century collections of ethnographic material culture took their inspiration. At the Great Exposition all of humanity's technical achievements were to be assembled under one roof – one universal and fully encompassing schema which excluded no one and not one thing from its purview. More significantly it was intended to edify and instruct the visiting public – provide them with a view of universal order, prosperity and progress which no theology up to this point had ever been able to do to such a telling degree. Thus the Great Exhibition served – to follow Pomian – as a window onto a universal realm of progress and prosperity just within everyone's reach, especially the inhabitants of the capital of the British Empire. The items on display became, using Pomian's term, semiophores – objects which do not have, or no longer have, a general practical use '. . . but which, being endowed with meaning, represented the invisible' (Pomian 1990) – the promise of a world of universal progress. Pitt-Rivers's famous and foundational collection for anthropology was first inspired by his visit to the Great Exhibition (Chapman 1985: 16). Even though something as ostensibly exotic as a neolithic axe found in Britain or an Aboriginal spear seemed to be as far removed as possible from the latest technical triumphs of nineteenth century industrialization, they all served together to emphasize a political, intellectual and cultural project based on empiricism, progress and perfectible unlineal evolution. As much as this justified European superiority, it also insisted on the perfectibility of all peoples (under European guidance) within the tradition of Enlightenment era liberalism and the universality of Man. The legacy of this impulse is still very much with us as rising nation states and creative ethnic self-determination assert claims towards inclusion and modernity, as Rowlands so cogently discusses in his contribution here.

There is a social reformist agenda here, which is often overlooked. These exhibitions not only brought in 'primitive' peoples within the unversalizing schema of European thought, but also brought in and edified the less enlightened in their own societies, serving as a vehicle for social reform. Both 'savage' and 'proletariat' were meant to be enlightened, edified and stimulated towards social progress and reform through these displays. Pitt-Rivers exhibited his collection in the severely deprived London working-class district of Bethnal Green in the 1870s with the purpose of edifying the masses so that they might more effectively participate in governance (Chapman 1985: 39). Eventually he realized the edifying purpose of his collection on the British working classes by setting up his collection on his Farnham estate. Thus, within this scheme of things, 'material culture' a peculiar super-category of objects was constituted and materialized as such. The highly contrived means by which some objects were separated out, and materialized as 'artefacts' within a 'material culture' in the aid of scholarship, colonial administration, museology and popular exoticism all served in their different ways to constitute and thereby materialize a very new, original and 'modern' category of objects.

These collections had a direct affinity with the rise in the nineteenth century of shopping and consumerism (Belk 2001). The exposition and the department store emerged at the same time and like the exposition, the newly invented institution of the department store with its vitrines and window displays provided views onto a desirable and more readily achievable (that is consumable) world that was just within reach – more universal and more open to a wider range of people than anything which preceded it. Such displays of material culture also served well within university settings as primary authoritative vehicles with which to peer into and understand non-western, past cultures and rapidly disappearing local 'folk' cultures. Anthropology appointments were often within museums and these museum collections formed the basis of instruction serving as the primary 'text', if you like, of early ethnographic and archaeological training (Hodder 1983:13). Similarly, In the United States, before the First World War, jobs in anthropology were in museums or research bureaux. Early teaching posts were split between museums and the first academic departments in anthropology at Harvard, California, and Pennsylvania. Franz Boas, himself worked jointly at the Anthropology Department of Columbia and the American Museum of Natural History (Fenton 1974: 19).

However, this super-artefact/intellectual tool was soon to lose its usefulness amongst Anglophone anthropologists and was virtually abandoned with the rise of British social anthropology, which sought to question the utility of these 'primary texts'. Rather than learning from these 'semiophores', this new turn, signalled most notably by Malinowski, amongst others, sought to understand societies directly through the innovative technique of participant observation over long extended periods of time: interview, discourse, observation and the reconstruction of social structure prevailed as a more perfect means of understanding. How societies functioned as social systems was more significant than how they could be placed within a schema of unlineal evolution based on material traits; the kinship diagram prevailed over the material culture 'fossil'. Transitional figures such as Boas became disillusioned with museum based work, becoming more interested instead in the social process which structured material culture (Boas 1907). The end result was an emphasis on kinship and social structure, the cornerstones of twentieth century social anthropology. Material culture as an intellectual and political tool became irrelevant, and faded by the wayside.

However, material culture retained its usefulness in other ways; most notably for its ability to materialize national identity in the creation of nationhood, as Rowlands and Bender illustrate here. Thus a number of traditions of material culture within folkloric studies remained and continued its nineteenth century mandate for delineating, materializing and stimulating social reform. The establishment of the Soviet Union witnessed the extraordinary institutionalization of the subject. The nineteenth century reformist and progressivist impulse was very much in evidence here on a scale Pitt-Rivers could never have dreamt of when Lenin 'abolished' archaeology as a 'bourgeois' science and re-created it as the study of the history of material culture in 1919. The evolution of the understanding of material culture in Soviet Russia probably followed more completely the logic of the world's fairs, serving as a new revolutionary form of social reform. This was an intellectual and political transformation that reworked archaeology not as an antiquarian discipline but as that branch of the human sciences which studies the 'history of material culture'. As such it was firmly part of history faculties, and a very long way from its earlier geological origins. Thus there was an understanding of a distinct field of material culture studies. Archaeology had the mandate to discern the processes of social change and progress through the study of material culture in the past, while more recent periods and the present were covered by the ethnologist, as part of a complementary analysis of the material basis of social evolution and progress. Archaeology and ethnology along with other arts and sciences were united in the common goal of the reconstruction of society according to Marxian principles towards the realization of communism and the social promise of modernity and progress – going some considerable way from Tylor's observation that the study of material culture ought to 'affect the courses of future opinions and institutions' (Ratzel 1896: xi). The Enlightenment era heritage found new impetus with the Russian Revolution. Marxian concepts of technical progress and unilineal evolution privileged the material world to an unprecedented extent and material culture studies served as revolutionary semiophores opening up on a new realm of social being. Within this context, the super-category had an extraordinary task to perform – to identify, chart and restructure social life towards a new future.

Melancholy and the Material

Within a climate stressing progress and social change, another element stood out that characterized the emerging preoccupation with artefacts and its study as material culture. This is a melancholic turn in the face of rapid social change both within the European imperial and national heartlands and colonial peripheries (see Rowland's introduction here). Traditional European life was changing quickly and much of traditional society, especially rural agrarian society was vanishing with the onslaught of industrialism. This nostalgia was a key element in the creation of foundational myths of industrialized nations. Similarly colonized societies that rarely had any contact with Europeans were rapidly changing with the expansion of colonial administration, trade and contact. This melancholy so well documented by writers such as Proust and Zola in France, was keenly felt by the early pioneers of anthropological research such as Haddon, Boas and others who were desperately aware of the precipitous rate of change in non-European societies whose ways of life were radically changing and whose traditional cultures were rapidly disappearing with the onslaught of imperial expansion in trade and administration.

This underlying concern with loss rather than consumption is probably the deeper motivation within material culture studies – rather than a view onto a world that is barely imaginable or about to come, this is a melancholic receding view. Material culture has been often, and rather uncritically, referred to as a mirror: 'As material culture, tools are the final objectification of intrinsic hopes. As imprinted thought and as engraved behavior, material culture becomes a mirror in which man can see himself' (Richardson, 1974: 12). However, this passive understanding reveals an unintended but important quality of material culture, that this view, either forwards or backwards, is constitutive and interpellative. Through its material constitution and the reiterative effects of its culturally produced durability, it becomes constitutive of desired and imagined subjectivies either nostalgic, futuristic or transformative which at times can have devastating consequences as Rowlands here describes in the context of India and the Former Yugoslavia. 'Cultural property' as constituted within material culture studies becomes the currency whereby nationhood or ethnic self-determination is ascribed according to how much of it one can show as 'proof' of one's coherence, integrity and worth. Rowlands points out little has changed since the Victorians – the emphasis on property and attendant notions of copyright is something both Rowlands and Küchler discuss here in greater detail, particularly Küchler in relation to the 'promiscuous' qualities of artefacts in the face of textual metaphors used to understand indigenous 'art' and its inhibition of material culture's promiscuity. Both Küchler and Pinney assert a renewed engagement with the nondiscursive, that is the phenomenological and somatic effects of material culture beyond textuality as does Tilley – a reassertion of the problematic relations between bodies and things (see Merleau-Ponty 1962 and Latour 1999), returning to a Maussian understanding of this fluid and hybrid relationship as revealed through Alfred Gell's *Art and Agency* (Gell 1998).

The Disillusionment of Objects and their Revival

Obviously the demise of unilinealism with the beginning of British social anthropology in the early twentieth century saw the demise of these objects as 'fossils' (which had served their purpose as appropriate semiophores). Material culture as cultural work was rendered increasingly useless in light of the developments of British social anthropology. In Britain few figures retained an interest. Wissler, a key figure in the field in the United States, already lamented the drop-off of interest before the First World War (Fenton 1974:20). Whereas Sayce in Britain was one of the few anthropological figures who pursued research in the inter-war period (Sayce 1933). Similarly the foundation of the department of anthropology at University College, London by Darryl Forde retained a link with this tradition through

Forde's interests in 'primitive technology' (Forde 1934) and a synthetic approach to the study of human society/evolution. Forde himself completed his Ph.D. at University College London but he was closely associated with Boas's students Kroeber and Lowie at Berkeley. He established and maintained the tradition of a teaching collection of material culture at UCL. However, by 1930 according to William Sturtevant, museum anthropology and traditional material culture studies reached their peak in most Euro-American traditions (Stocking 1985: 9). Between the wars and afterwards, material culture studies become increasingly irrelevant and stagnant.

The Anglophone tradition of material culture studies receded but did not disappear. The renewed force it was to achieve in the second half of the twentieth century drew much of its force in Britain from the pioneering work of the archaeologist Gordon Childe which reasserted the social within archaeological studies along with later Marxist-inspired critiques of consumerism in sociology. Childe was a keen observer of Soviet archaeology with its emphasis on the socially diagnostic aspect of material culture studies. Childe's subtle Marxism, his rejection of antiquarianism and assertion of the significance of social processes in the study of archaeology paved the way along with others for the eventual formation of the 'New Archaeology' in Britain and America, which broke the slumbering theoretical innocence of antiquarian Anglo-American archaeology with a renewed emphasis on the study of social processes as materialized in the material culture of the past.

Material culture studies as we come to know it now, emerged within the British tradition and gained renewed impetus amongst a group of variously Marxist-inspired archaeologists based at The Archaeology Institute of University College London and the archaeology department of the Faculty of Archaeology and Anthropology at Cambridge University. Many of the figures associated with this period are represented in this volume: Bender, Miller, Rowlands who worked closely with Jonathan Friedman, and Tilley. At the same time at the UCL department of anthropology in the 1970s, Mary Douglas was working on some of the key texts that brought this field of interest back into prominence: *Purity and Danger* and, with Baron Isherwood, *The World of Goods*. Her work was paralleled by that of other anthropologists such as Arjun Appadurai and Pierre Bourdieu. French trends particularly within the work of the Annales School of History took significant turns in reconsidering the significance of material cultures studies (Braudel 1992, Baudrillard 1996, Barthes 1973). As Miller here notes, the emerging significance of semiology from within linguistics (Barthes 1973, Baudrillard 1996) and especially structuralism (Lévi-Strauss) saw the re-evaluation of the material within symbolic systems. This is especially so in Lévi-Strauss's work on Pacific Northwest Coast masks (Lévi-Strauss 1988) and his reappraisal of the work of Franz Boas. Structural Marxism provided a powerful critique of the role of objects in symbolic systems and social structures at the time of the social tumult and student riots of the 1960s and early 1970s. These were powerful conceptual tools with which to confront post-war capitalist countries with critical and diagnostic Marxist studies of material culture.

The rise of interest in semiotics and structuralism had an important effect on the revival of interest in material culture studies in the United States as well (Glassie 1975, Deetz 1977), as did offshoots of the 'New Archaeology' through the development of ethno-archaeology and the resulting interest in modern material culture studies on their own as in the works of Rathje, Schiffer and later Marxist archaeologists such as Layton, Paynter and Leone. With the 1980s of course, this direction had by no means disappeared, but a new reappraisal emerged which began to see consumption as an active process, whereby individuals actively appropriated material goods towards the creation of inalienable culture (Miller 1987, Belk 2001). Finally, the *Journal of Material Culture* was established in 1996 at University College London (Editorial 1996) being the first Anglophone academic journal explicitly dedicated to the interdisciplinary study of the field and which has been edited at various times by the individuals represented in this volume.

Within the British Tradition, the post-war revival of interest in material culture studies is often associated with the Curl Lecture by Peter Ucko on the cross-cultural study of penis sheaths (Ucko 1969). Ucko following in the footsteps of Darryl Forde played a key role in formalizing the Material Culture group at UCL. That the re-emphasis on the material should have been heralded by the arch material signifier, the sheathed phallus, whether intentional or not, is a point best explored by Lacanian analysts. Regardless it is a fundamentally apt beginning for a reappraisal of the presence of material signifiers in anthropological analyses. The foundational Curl lecture, however, raises some interesting questions regarding a masculinist bias in material culture studies that has been rarely discussed. Feminist analyses have shown that material presence (which is what material culture studies deal with: the socially constituted and materialized physical artefact) is a consequence of a deeply placed masculinist bias – as feminine subjectivities are understood in terms of their inherent 'lack' vis-à-vis the elemental presenced material signifier, the phallus (Butler 1993). Ruth Olendziel's discussion of technology, culture and gender (Oldenziel 1996) explores another masculinist bias in material culture studies: the link to industrial modernization, progress and imperial governance, and its overt emphasis on the material and production (male) at the expense of use and consumption (female): a focus which does not emerge openly until the anthropological studies of consumption in the 1980s (Miller 1987, Douglas and Isherwood 1979, Appadurai 1986). As Judith Butler has suggested, much of the materialized world is forged within this masculinist bias, that sees the realm of the 'feminine' as one of lack, and constitutive of the 'masculine signifier', thus forming what she refers to as the 'constitutive outside' that defines and materializes the dominant 'masculine'. Thus the 'feminine', an abject category, is unmaterialized in two senses, as not 'mattered physically' and also as not 'mattering' as social worth – its absence thereby secures and delineates the contours of the 'masculine'. So if this emphasis on materiality that presences the masculinist signifier renders our understanding of the feminine and women problematic, what does it do for other subjectivities and other states of being? Olendziel is one of the few voices to call into question the universality of the concept of material culture itself despite the prevalent celebration of its universality which to this day is still triumphantly announced. As a universal, this may be just an empty sign, but it is a sign nonetheless that constitutes a bracketing and certain original exclusions that the history of this topic of study demonstrates (see Löfgren 1997) as have the masculinist universals revealed by various traditions of feminist scholarship. Olendziel argues quite rightly that this signifier, though empty, has an ideological basis that might not permit us to understand those processes that are entailed in materialization and the exclusions that inevitably result.

The Present

Material culture as we understand it is a direct consequence of the collecting traditions of the nineteenth century, liberal Enlightenment era notions of universality, colonial expansion, industrialization and the birth of consumerism. As stated before, these collections were the primary means by which we studied other societies in distant time and space. We abandoned these studies to the promises made by social anthropology, which sought to go direct to the source rather than try and understand and translate it through ethnographic collections. If we consider Krystoff Pomian's thesis here, these earlier ethnographic collections were clearly attempts to mediate between two worlds, one known (Western) and one not known and invisible (non-Western), that could be comprehended through these mediating objects we call material culture. There is an element of sacrifice and wastage here in terms of utility not unlike the negation of the feminine as 'lack' – as that which is precluded or 'pre-disposed', to borrow Strathern's useful term (Strathern 2001), to ensure a desired category. However, Bataille might be instructive here in his similar investigations of the Potlatch and other practices within what he describes as sacrificial economies. These are sacrifices of objects, attempts to render ultimate inalienability be they through the creative destruction of archaeological sites, or the deathlike still-life artefact assemblages of museum collections. Through this inalienability, ideal worlds and states of being are delineated, whether it be the small sacrifices a housewife makes in her shopping excursions to realize a familial ideal (Miller 1998b) or the grandiloquent sacrifices of previously useful objects, as in Potlach rituals or those in museums – objects are withdrawn from one sphere of social use, wasted in relation to that sphere to constitute and materialize alternate ideal realms. As suggested earlier, material culture studies as part of a sacrificial economy has historically occurred within a framework of social purpose, which required the constitution of such super-material objects – material culture – to facilitate these goals whether industrial progress, social revolution or critical consciousness.

Daniel Miller has noted that the study of material culture is an integrative endeavour (Miller 1983). Thus one might hazard to describe here three attempts where material culture has emerged as an integrative intellectual project: evolutionary thought in the nineteenth century; Marxian social analysis and revolution in the early twentieth century and progressivist New Archaeology and Marxian social theory in the second half of the twentieth century. The problem with current approaches is the lack of an overtly integrative intellectual project, a consequence of the postmodern condition and the demise of Enlightenment era ideologies such as liberal notions of universalism, progress and Marxism. The fragmentation of such narratives that otherwise describe our so-called 'postmodern' condition may in part explain material culture studies' persistent and increasing heterogeneity as it surfaces within so many disciplines. Its instability is a consequence of its virtue – being a socially motivated and contingent materialization of objects into systems of material culture. It has never really been a discipline – it is effectively an intervention within and between disciplines; translations from one realm into another. But it is precisely this persistent heterogeneity and the proven ability of material culture studies to translate (by virtue of its disruptive abilities) not just simply between different and incommensurable social and physical realms, but between disciplinary realms as both Rowlands and Bender argue here. This might partially explain the increasing turn towards the material across the various disciplines of the humanities in addition to the consequences of the rapidity of culture change which typically evinces a melancholic preoccupation with the material as a means of coping with change. The moment we are in right now is just one in a history of many other attempts to focus and mediate between a realm rapidly becoming invisible and unrecognizable from our own. The nineteenth century idea that culture change could be evinced from our relationship to objects and thereby coped with more effectively has not really shifted much.

The reconstitutive (and destructive) operation of material culture involves a certain degree of waste and sacrifice; with war as the most spectacular expression of 'the transformation of matter through the agency of destruction' (Saunders, this volume). It also transforms a mostly inarticulate realm of sensual experience into the two dimensions of a scholarly text or the '*nature-morte*' of the museum display (as in all translations something is always 'lost'). This suggests a decrease in physicality across dimensions – moving sensual reality increasingly towards the dimensionless and ephemeral. Vast realms of sensual reality and utility are removed, transformed and made into the sensually 'dead' objectifications of 'material culture' we call an ethnographic monograph which preclude as required by a modernist science the more promiscuous and multiple meanings generated by the materialized 'artefact'. So much, and quite necessarily so, is wasted in terms of twentieth century cultural work – the troublesome fetish of a conservative Marxist discourse is suppressed, rendered harmless and erased by edifying analyses that attempt to keep the transfixing , enchanting and promiscuous affects of the artefact at bay (Belk 2001; Editorial 1996). This process is bemoaned by Löfgren as we neglect and unproblematize the materiality of material culture. We no longer dare to stroke those 'consenting molecules' (Löfgren 1997) which constitute material culture as our antiquarian ancestors did. The erotics and attendant politics of this materiality are inadequately discussed. There is a promiscuity here as both Pinney and Küchler describe that is rarely explored (but see Shanks 1992) and hindered by our preoccupation with textuality. Most of our publications deny us any visual representation of the very physical objects we explore. This was never the case in the beautifully illustrated discussions of material culture in the past and their exquisite display when the affects of these objects were at their most problematic from the standpoint of mid-twentieth century anti-consumerist and post-colonial anxieties. Their visuality and form was the primary vehicle of authority and information, the text was merely supplementary and discursive (Lucas 2001, see also Thomas 1997: 93–132). This is the reverse of how we recently have valued the authority of such visual materialization of material culture. That we have sanitized them to such a degree, evacuating them into inaccessible collections, constituting them as edifying discursive texts, and at times even rendering them dangerous – as some frustrated Native American groups have found their repatriated objects conserved with highly poisonous substances. Conservation is anything but that: it is a very active and deliberate process of materialization; it 'conserves' nothing but 'produces' everything, as we can learn from Bruno Latour's work (Latour 1999). So what are the social effects and costs of such productive materializations such as 'conservation'? Are these poisoned artefacts the result of some misplaced fear of the seduction of the commodity fetish – a legacy of a conservative critique that sought to deny earthly seductions in an effort to achieve an idealized order – or something else entirely?

Waste, Change and Ephemerality

Material culture studies has been described by Rowlands as an intellectual refuge '. . . during periods of antipathy when anthropology's rupture with its nineteenth century origins threatened to abolish all questions that recalled a tainted past. It may always need to be preserved as such, since to rupture such a category is always to place its contents in danger' (Rowlands 1983: 16). This emphasis on translation and rupture suggest a different perspective from the imperial, universalistic, panoptic one of the nineteenth century. The issue of translatability from one realm to the another, the invisible into the visible, described by Pomian recalls a recent point made by Judith Butler regarding left politics as being one centred on translation, from an interstitial position 'to shatter the confidence of dominance, to show how equivocal its claims to universality are, and, from that equivocation, track the break-up of its regime, an opening towards alternative versions of universality that are wrought from the work of translation itself' (Butler, Laclau and Žižek, 2000: 179).

The interstitial positions occupied by material culture studies provide a platform for a critical engagement with materiality for understanding issues facing us such as the fluidity of gender and body/object interfaces, recyclia, biotech, genetic engineering and the Internet – in short, those key materializing and transformative processes that shape new inclusions and exclusions as the critical focus of material culture studies such as new kinds of bodies, forms of 'nature' and political subjects.

One might consider here the nature of alienability as a tendency towards fluidity that denies a certain 'cultural' mass. As this fluidity quickens it moves, losing the 'weight of tradition' towards an increasingly 'lighter' and immaterial state (Oldenziel 1996: 63). This process is like that described by Thompson whereby objects are literally transformed in terms of their physicality and durability as a consequence of the cultural work that transforms rubbish on its way to becoming immaterial dust into durable artefacts – materiality is by no means a non-negotiable and unquestionable empirical reality it is a produced social one. As Thompson states 'Those people near the top have the power to make things durable and to make things transient . . .' (Thompson 1994: 271). This socially produced durability is the effect of extensive cultural interventions – the exchange value of the market or the science and politics of museum curation being prominent amongst others. In short this is the production of what one might call an artefactual effect (see Fletcher 1997a, b); the result of a profound social alchemy. This massiveness, or this so-called 'weight of tradition' is shown by Gilles Lipovetsky (Lipovetsky 1994) to be entirely undermined by the ephemerality of the fashion system of consumerism dating back to the nineteenth century, that since this time has actively worn away at the 'gravity' and 'mass' of custom. The crushing ephemerality of late capitalism (its constant material flux), its 'tragic lightness' (as Lipovetksy calls it) combined with the increasingly immaterial nature by which individuals assert agency and intervene in the social world (information technology over production, the Internet, the extreme mobility and liquidity of capital, and the intense rate of consumption and waste production) all create a situation where the insistence on the peculiar, limited and highly contingent fixity of the material artefact seems all the more inadequate to cope with the social effects of these increasingly ephemeral, highly fluid and immaterial interventions within the material world that sustains us.

This issue is becoming more the focus of recent work in material culture studies that focuses on cultures of waste, destruction (Saunders, this volume), recycling, divestment, moving, capital flows, etc., which suggests that the processes of materialization are more significant than materiality itself and in fact variably constitutive of it – material culture itself is just a peculiar moment in these processes – an alchemical cultural effect which serves as a diagnostic formed by processes of waste and sacrifice required of our various cultural projects. This more recent work on materiality and material culture has focused on a certain critical empiricism (Miller personal communication, Buchli and Lucas 2000, Oldenziel 1996: 66) which examines closely the terms by which discursive empirical reality is materialized and produced. This is a continuation of the suggestions of Bataille which moved the focus of consumption and the understanding of material culture from consumption and use value, to an exploration of the processes of waste, and the logics of sacrificial economies rather than normative notions of utility. This we can understand as a preoccupation with the means by which alienability occurs, how things are released, given away, wasted, taken away, sacrificed or disposed of towards the creation of the social terms of existence. These are key concerns within recent studies of recycling and moving and similarly the repatriation of artefacts and reburial of remains. These are all actions of one sort or another that facilitate a transformation of the materiality of material culture in terms of durability and visibility. New subjectivities are facilitated through this process which tends to diminish the materiality of material culture and even to move out of the realm of durable 'conserved' material culture itself. In the case of recent repatriation and reburial controversies – what for a museum curator represents an almost iconoclastic wastage of precious artefacts (a fact that is undeniable from the point of view of orthodox Western science) is on the other hand the highly creative act of cultural construction and consolidation from the point of view of some indigenous groups – and additionally, a radical reconstitution of identity facilitated by the very same objects of material culture that facilitated the original exclusions and subaltern status of such groups in the first place (Jacknis 1996: 209). Conservation and creative destruction become problematic in the face of differing and conflicting material strategies vying for social control (see Rowlands and Saunders this volume).

The more recent emphasis in material culture studies, one might say, in many respects has been its most traditional – that is in terms of its focus on translation and the material processes at work to facilitate a view from one realm on to another. Such translations are more significant and more frequent in terms of the increasing rapid change and superfluity of knowledge and goods. This is a point explored by Lipovetsky in terms of the significance of the ephemerality of fashion as non-durable, changing, frequently wasted and fleeting to facilitate a view from one realm on to another. Rather than suggesting a lack of distinction, authenticity and inalienability – the inherent alienability of fashion, as fleeting, frequently cast off – are the very terms by which social viability and enfranchised subjectivity are possible. Through the democratizing and enfranchising effects of the fluid and mobile immateriality of fashion all stabilizing authority and tradition is challenged. A constantly fluid and immaterial means is established by which to assert new subjectivies. Neo-pragmatist thinkers such as Rorty and Radical Democratic theorists such as Laclau and Mouffe and Butler argue for the importance of this instability and openendedness that never lets any one particular way of getting things done ever get the upper hand. This is the ethical 'scrappiness' of Smith (1988) or the disorderly virtuous cities of Sennett, which are believed to best secure democratic freedoms (Sennett 1971). The production and waste of objects and their constitution and dissipation are the two sides of the larger processes of materialization that facilitate the terms of social life, perpetuating its inclusions and exclusions as well as reworking and challenging them.

Material culture's ability to constitute through the cultural articulation of its durability as increasing inalienablity shows that it has not disappeared in the present day and is still very much in force. Material culture functions as a means of resistance against globalization or as a way of countering colonial legacies (Rowlands this volume), or through consumption facilitates inalienable authenticity (Miller 1987), and the generation of various 'strategic essentialisms' – the 'cultural property' of Rowlands's contribution here. However, as various neo-pragmatist and Radical Democratic thinkers have suggested, these critical interventions are momentary, contingent and strategic – creating what one might call a critical empiricism. These concerns echo Lipovetsky's understanding of the fashion system as conspiring against the solidity of objects, and that personal liberty is in fact guaranteed by the increasing ephemerality of the material world. Its rapid flux does not allow one to make firm attachments either to an object or an ideology, or tradition. An ideology of superficiality within a rapidly changing and wasteful material world ensures that no one form of materiality will ever prevail or get the upper hand, which can be rejected and left behind like the poetic metaphors of Richard Rorty: 'The proper honor to pay new, vibrantly alive metaphors, is to help them become dead metaphors as quickly as possible , to rapidly reduce them to the tools of social progress' (Rorty 1991: 17). What is more important probably is not to study the materializations themselves but rather what was wasted towards these rapid and increasingly ephemeral materializations (what Strathern refers to as the universe of meanings predisposed by social conventions (Strathern 2001)) The realm of the abject, the realm of the wasted beyond the constitutive outsides of social reality is where critical work needs to be done (rubbish studies, divestment studies, the disenfranchised of globalization, the 'non-places' of Augé (Augé 1999) and the general effects of late capitalist ephemerality). This is the territory of 'tragic lightness' described by Lipovetsky. The ephemerality of human interaction, the inability of any one regime to take hold subject to the ever-increasing individualized needs of consumerist novelty means that ontological security is tentative and supremely contingent at best. This ethos of ethical disorder which ensures that no one regime gets the upper hand and the boundaries of social legitimacy can always be challenged means that even though 'The consummate reign of fashion pacifies social conflict; it allows more individual freedom, but it generates greater malaise of living [. . .] which renders us increasingly problematic to ourselves and others' (Lipovetsky 1994: 241). How people negotiate the increasing immateriality and alienability of our material world is one of the challenges facing material cultures studies.

The fragmented nature of the discipline is hardly a sign of crisis, but rather a testimony to its vigour in an expanded and diffuse realm of social inquiry. Within this of course lies the issue of materiality, the various ways we materialize social being and the ways in which this process is challenged in light of rapid social change and the increasingly ephemeral nature of our social interactions. Under such circumstances numerous voices disappear as quickly as they appear, or are never able to appear at all, buried within the rapid superfluity of information and materiality. How things come to matter both physically and socially, how the terms of materiality are reconfigured to facilitate various forms of social inclusion and exclusion are questions which become increasingly relevant. This is another way of understanding materiality not so much as physics but as cultural process – the immateriality of cyberspace can cause as much pain (Haraway 1991) because of the social effects by which these materialities or immaterialities are constituted. The material realm has not been supplanted, the virtual realm works alongside in a hybrid fashion to facilitate such connections, views and realms as most innovations in the past have done (see Haraway 1991 and Latour 1999). Its 'artefactuality' (Fletcher 1997a) is just as effective as it was early on: the Internet as much as the constituted and 'conserved' artefact, or nineteenth century engraving are different constitutive representations. They have specific social effects as relevant along the continuum of various materialized and de-materialized states from the actual object to its manifestation in cyberspace. They all produce a certain artefactuality (Fletcher 1997b) – that is an artefact effect with contingent social purpose: the 'factishes' of Latour (1999). In this respect anthropology since its beginnings has always traditionally dealt with and produced the virtual – whose respective social worths are assessed in terms of how they are able to mediate between one state and another with their respective social effects. What is very different is how we consider and configure the material conditions of our interactions, that is how does materiality function, what does it do, what are its new social costs and who is included or excluded, given a voice or silenced. A number of the contributions of this volume provide excellent examples such as the conflicts over cultural properties in India mentioned by Rowlands, or those over Stonehenge described by Bender and the kinds of subjectivities that could be accommodated within the changing Soviet home described here by Buchli. In a sense, looking at what happens before and after the artefact is more significant than the artefact itself; that is, the terms of materiality rather than material culture itself and the differential ability of individuals to participate in these processes is more important. As Butler has suggested in relation to the materiality of gender, this means '. . . a return to the notion of matter, not as site or surface, but as *a process of materialization that stabilizes over time to produce the effect of boundary, fixity, and surface we call matter*' (Butler 1993: 9, see also Strathern 1988). The materializing function of archaeological and anthropological projects in material culture studies serves to render discursively legible, groups, worlds, individuals, subjectivities and experiences that were otherwise outside of the discursive realm (Buchli and Lucas 2001), thus they help refigure the boundaries of inclusion – suggesting possible worlds and views that are increasingly silenced, overlooked and forgotten in the increasingly ephemeralised world of human interaction – and thereby address and challenge the social and ontological costs of this 'tragic lightness' (Lipovetsky) which surrounds us.

References

Appadurai, A. (1986), *The Social Life of Things: Commodities in Cultural Perspective*, Cambridge: Cambridge University Press.

Attfield, J. (2000), *Wild Things: The Material Culture of Everyday Life*, Oxford: Berg.

Augé, M. (1999), *Non-Places: Introduction to an Anthropology of Supermodernity*, London: Verso.

Barthes, R. (1973), *Mythologies*, St. Albans: Paladin.

Bataille, G. (1991), *The Accursed Share vol. 1*, New York: Zone Books.

Baudrillard, J. (1996), *The System of Objects*, London: Verso.

Belk, R.W. (2001), *Collecting in a Consumer Society*, London: Routledge.

Boas, F. (1907), 'Some Principles of Museum Administration', *Science* 25: 921–33.

—— (1955,) *Primitive Art*, New York: Dover Publications.

Braudel, F. (1992), *Civilization and Capitalism*, Berkeley: University of California Press.

Buchli, V. and Lucas, G. (2001), *Archaeologies of the Contemporary Past*, London: Routledge.

Butler, J. (1993), *Bodies that Matter*, London: Routledge.

Butler, J., Laclau, E. and Žižek, S. (2000), *Contingency, Hegemony, Universality: Contemporary Dialogues on The Left*, London: Verso.

Chapman, W.R. (1985), 'Arranging Ethnology: A.H.L. F. Pitt-Rivers and the Typological Tradition', in G.W. Stocking Jr (ed.), *Objects and Others: Essays on Museums and Material Culture*, Madison: University of Wisconsin Press.

Deetz, J. (1977), *In Small Things Forgotten*, New York: Anchor Press.

Douglas, M. and Isherwood, B. (1979), *The World of Goods: Towards an Anthropology of Consumption*, London: Routledge.

Editorial (1996), *Journal of Material Culture* 1(1): 5–14.

Fenton, W.F. (1974) 'The Advancement of Material Culture Studies in Modern Anthropological Research', in M. Richardson (ed.), *The Human Mirror*, Baton Rouge: Louisiana State University Press.

Fletcher, G. (1997a), 'Excavating Posts: an Archaeology of Cyberspace', Paper presented at *WIP-ing post* Conference, University of Queensland, Australia.

Fletcher, G. (1997b) Excavating the Social. Paper presented at *Rethinking the Social* Conference, Griffith University, Queensland, Australia.

Forde, D. (1934), *Habitat, Economy and Society*, London: Methuen & Co. Ltd.

Forty, A. and Küchler, S. (eds) (1999), *The Art of Forgetting*. Oxford: Berg.

Gell, A. (1998), *Art and Agency: an Anthropological Theory*, Oxford: Clarendon.

Glassie, H. (1975), *Folk Housing in Middle Virginia*, Knoxville: The University of Tennessee Press.

Haddon, A.C. (ed.) (1935), *Reports of the Cambridge Anthropological Expedition to Torres Straits*, vol. 1. Cambridge: Cambridge University Press.

Haraway, D. (1991), *Simians, Cyborgs, and Women: The Reinvention of Nature*, London: Free Association Books.

Hodder, I. (1983), Material Culture Studies at British Universities: Cambridge, in *Things ain't what they used to be*, Daniel Miller (ed.) Royal Anthropological Institute News, 59: 13–14.

Jacknis, I. (1996), 'The Ethnographic Object and the Object of Ethnology in the Early Career of Franz Boas', in W. Stocking Jr (ed.), *Volksgeist as Method and Ethic: Essays on Boasian Ethnography and the German Anthropological Tradition*, Madison: University of Wisconsin Press.

Kingery, W.D. (ed.) (1996), *Learning From Things: Method and Theory in Material Culture Studies*, Washington DC: Smithsonian Institution Press.

Latour, B. (1999), *Pandora's Hope: Essays on the Reality of Science Studies*, Cambridge, MA: Harvard University Press.

Lévi-Strauss, C. (1988), *The Way of the Masks*, Seattle: University of Washington Press.

Lipovetsky, G. (1994), *The Empire of Fashion: Dressing Modern Democracy*, Princeton: Princeton University Press.

Löfgren, O. (1997), 'Scenes from a Troubled Marriage: Swedish Ethnology and Material Culture Studies', *Journal of Material Culture*, 2 (1): 95–113.

Lowie, R.H. (1960), *The History of Ethnological Theory*, New York: Holt, Rinehart and Winston.

Lucas, G. (2001), *Critical Approaches to Fieldwork: Contemporary and Historical Archaeological Practice*, London: Routledge.

Marx, K. (1986), *Karl Marx: a Reader*, J. Elstner (ed.) Cambridge: Cambridge University Press.

Merleau-Ponty, M. (1962), *The Phenomenology of Perception*, London: Routledge.

Miller, D. (1983), 'Things ain't what they used to be', *Royal Anthropological Institute News*, 59:5–7.

—— (1987), *Material Culture and Mass Consumption*, Oxford: Basil Blackwell.

—— (1998a), *Material Cultures: Why Some Things Matter*, London: University College London Press.

—— (1998b), *A Theory of Shopping*, Cambridge: Polity.

Morgan, L.H. (1978), *Ancient Society*, New York: Labor Press.

Mouffe, C. (1993), *The Return of the Political*, London: Verso.

Mouffe, C. (1996), 'Deconstruction, Pragmatism and the Politics of Democracy', in C. Mouffe (ed.), *Deconstruction and Pragmatism*, London: Routledge.

Oldenziel, R. (1996), 'Object/ions: Technology, Culture, and Gender', in W.D. Kingery (ed.), *Learning From Things: Method and Theory in Material Culture Studies*, Washington DC: Smithsonian Institution Press.

Pearce, S. (1994), *Interpreting Objects and Collections,* London: Routledge.

Pomian, K. (1990), *Collectors and Curiosities: Paris and Venice, 1500–1800*, Cambridge: Polity Press.

Rabinow, P. (ed.) (1984), *The Foucault Reader*, London: Penguin Books.

Ratzel, F. (1896), *The History of Mankind*, London: Macmillan and Co. Ltd.

Richardson, M. (ed.) (1974), *The Human Mirror*, Baton Rouge: Louisiana State University Press.

Rorty, R. (1991), *Essays on Heidegger and Others*, Cambridge: Cambridge University Press.

Rowlands, M. (1983), University College London, in *Things ain't what they used to be*, Daniel Miller (ed.) Royal Anthropological Institute News, 59: 15–16.

Sayce, R.U. (1933), *Primitive Arts and Crafts*, Cambridge: Cambridge University Press.

Schnapp, A. (1996), *The Discovery of the Past: the Origins of Archaeology*, London: British Museum Press.

Sennett, R. (1971), *Uses of Disorder: Personal Identity and City Life.* London: Allen Lane.

Shanks, M. (1992), *Experiencing the Past: On the Character of Archaeology*, London: Routledge.

Smith, B.H. (1988), *Contingencies of Value*, Cambridge, MA: Harvard University Press.

Steadman, P. (1979), *The Evolution of Designs: Biological Analogy in Architecture and the Applied Arts*, Cambridge: Cambridge University Press.

Stocking, G. (ed.) (1985), *Objects and Others: Essays on Museums and Material Culture*, Madison: University of Wisconsin Press.

Strathern, M. (1988), *The Gender of the Gift*, Berkeley: University of California Press.

—— (2001), 'The Aesthetics of Substance', in N. Cummings and M. Lewandowska (eds) *Capital*, London: Tate Publishing.

Thomas, N. (1991), *Entangled Objects: Exchange, Material Culture, and Colonialism in the Pacific*, Cambridge, MA: Harvard University Press.

Thomas, N. (1997), *In Oceania: Visions, Artifacts, Histories*, Durham, NC: Duke University Press.

Thompson, M. (1994), 'The Filth in the Way', in S. Pearce (ed.), *Interpreting Objects and Collections*, London: Routledge.

Tilley, C. (1991), *Material Culture and Text: The Art of Ambiguity*, London: Routledge.

Trigger, B. (1989), *A History of Archaeological Thought*, Cambridge: Cambridge University Press.

Ucko, P. (1969), 'Penis Sheaths: a Comparative Study', *Proceedings of the RAI*, pp. 24–66.

–2–

Metaphor, Materiality and Interpretation
Christopher Tilley

Introduction

During the past thirty years some of the most exciting and innovatory ethnographic and archaeological studies of material culture have exploited analogies with language to provide a fresh way of understanding of what things mean, and why they are important. Structuralist approaches have led us to think about things as communicating meaning like a language, silent 'grammars' of artefact forms such as sequences of designs on calabashes, pots or bark cloth. Similarly 'grammars' of household and village space, gravegoods and burials, etc. have been produced and then linked back to a structure of social and political relations in various ways (see Tilley 2000 for a recent review). Things have thus become regarded as texts, structured sign systems whose relationship with each other and the social world is to be decoded. In various post-structural approaches to material forms the metaphors of language, or discourse, and text have remained dominant in an understanding of things. The new emphasis here has been on polysemy, biographical, historical and cultural shifts in meaning, the active role or 'agency' of things in constituting rather than reflecting social realities, power/knowledge relations and the poetics and politics of the process of interpretation itself, that we write things rather than somehow passively read off their meanings independently of our social and political location, values and interests.

But a design is not a word and a house is not a text: words and things, discourses and material practices are fundamentally different. Clearly linguistic analogies may serve to obscure as much as they may illuminate the nature and meanings of things as material forms. Yet (at least as academics) we primarily have to write and speak of things, transform them into utterances and thus risk domesticating their difference from the language used to re-present them. Much as perhaps we might like it, the problem of language will not go away in the study of the things. It is only through the use of words that we can claim, assert, investigate and understand why things matter and why a study of them is important, why it makes a difference to an understanding of persons and their social worlds. It is this general problem of how we cope with language in the study of things that I attempted to explore in some detail in my book *Metaphor and Material Culture* (Tilley 1999) of which the study of canoes in Vanuatu, reproduced in this volume, forms a part.

To be human is to speak, to be human is also to make and use things. Neither language or the production, reception and use of material forms can be claimed to have any ontological primacy. As differing modes of communication the linguistic forms of words and the material forms of artefacts play complementary roles in social life. What links together language use and the use of things is that both arise as products of an embodied human mind, i.e. a mind that makes sense of and intervenes in the world through the sensuous and carnal capacities of the human body. Our flesh is a connective fabric of carnal tissue binding us to the world linking together words and things in the creation of meaning and the performative sphere of action. I argue that language use is thoroughly metaphorical in nature. Our speech is laden with metaphors because we think metaphorically. The material counterpoint to verbal metaphor is the solid metaphors objectified in the forms of artefacts. Metaphors and metonymy (part-whole relations) allow us to see similarity in difference and permit us to connect the world together. They thus can be said to constitute the corporeal flesh of our language and the material flesh of things. They doubly constitute our meaning and experience, providing a meeting ground between languages and discourses of representation, feeling, emotion and multiple experiential modes of engagement with the world.

Metaphor is a primary way in which persons and cultures make sense of the world. When we link things metaphorically we recognize similarity in difference, we think one thing in terms of the attributes of another. This position, which may be labelled a poetics of mind (Gibbs 1994) emphasizes that thought arises from our embodied experience. Hence many metaphors are grounded in the human body and in mental images of the world based on bodily experience. Such experiences and images are always mediated through social experience and thus are culturally variable. To cite just one example, the Dogon of Mali conceive the world as a gigantic human organism. The village is a person lying north-south, smithy at its head, shrines at its feet. The Dogon house is an anthropomorphic representation of a man lying on his side and procreating. There is an entire geology of the body. Different minerals correspond to different bodily organs. Rocks are bones, red ochre is blood. Words are likened to grain, speech to germination, divination to winnowing. Body parts have analogues in grain, the nose being likened to the germ (Griaule 1965). This is clearly a corporeal and sensuous (animistic and anthropomorphic) way of relating self and culture to the world. Metaphor and metonmy are situated in the practical activity of the Dogon in engaging with the land and the growing of grain, a participatory logic of practice.

Metaphors are both creative and infinitely generative in their allusions and the manner in which they permit the creation of meanings. They are not an embellishment or an elaboration of an originary and primary literal language (the traditional theory of metaphor going back to Aristotle) but constitute its very essence as a mode of communication. 'Dead' bodily metaphors are so ubiquitous and embedded in our own thought that we rarely realize that we are even using them when we speak (e.g. expressions such as the leg of a table, the face of a clock, I see [i.e. understand] what you mean). To be human is to think through metaphors and express these thoughts through linguistic utterances and objectify them in material forms. The essence of metaphor is to work from the known to the unknown, to make connections between things so as to understand them. A metaphorical logic is thus an analogic logic serving to map one domain in terms of another and differs fundamentally from the abstract and universal digital logic presupposed by 'classical' structuralist approaches. Whether we realize it or not, all the interpretative work we do in the social sciences has a metaphorical foundation. Metaphors are thus the very medium and outcome of our analysis.

The chapter about canoes reproduced here attempts to show how material or solid metaphor works through an extended analysis of one ubiquitous artefact form in the south Pacific. It is based on fieldwork undertaken on Wala island, Vanuatu in 1995 and was written in 1996. The primary sources of ethnographic inspiration are the works of Munn (1977; 1986), MacKenzie (1991) and Strathern (1988). A somewhat contrasting 'dialogic' approach to the material symbolism of the canoe can be found in a recent paper by Barlow and Lipset (1997). The main points I try to make here, in the exploration of the material metaphors of the canoe are as follows:

1. The canoe as ubiquitous utilitarian artefact links together all the fundamental domains of social life and as an artefact of tradition (*kastom*) the past and the present.
2. Through the process of fabricating and using canoes people make both themselves and their social relations.
3. Canoes silently allow people to talk about themselves and their social relations in ways impossible in verbal discourse. They permit the unsaid to be said with regard to clan and gender relations.
4. They are dynamic forms articulating space-time and mediating social ties. They link opposed metaphorical qualities of land and sea, maleness and femaleness, rootedness and journeying, at the heart of a Melanesian island conception of the world.
5. Their metaphorical meanings are multifarious, enabling them to combine and embrace contradictory principles and tendencies in social life. The power of the imagery resides in its condensation of reference linked with the sensual and tactile qualities of its material form and reference to the human body.
6. The form of the canoe and the metaphorical attributes associated with it permits the creation of vital referential 1inks to feasting, grade taking ceremonies, pig sacrifice and dance at the heart of social reproduction. Artefacts are good to think through their metaphorical relations with others. They can act as agents of reproduction or innovation according to the manner in which they are both appropriated and transmitted.

The overall aim of this chapter is to suggest that artefacts perform active metaphorical work in the world in a manner that words cannot. They have their own form of communicative agency. It follows that without an exploration of the meta-phorical powers of things and the effects that these things have on people's lives we cannot adequately know or understand ourselves or others, what makes up our identity and culture, past or future.

References

Barlow, K. and Lipset, D. (1997), 'Dialogics of material culture: male and female in Murik outrigger canoes', *American Ethnologist* 24(1): 4–36.

Gibbs, R. (1994), *The Poetics of Mind*, Cambridge: Cambridge University Press.

Griaule, M. (1965), *Conversations with Ogotemmeli*, Oxford: Oxford University Press.

MacKenzie, M. (1991), *Androgynous Objects*, Melbourne: Harwood Academic Press.

Munn, N. (1977), 'The spatiotemporal transformations of Gawa canoes', *Journal de la Société de Océanistes* 33: 39–53.

—— (1986), *The Fame of Gawa*, Cambridge: Cambridge University Press.

Strathern, M. (1988), *The Gender of the Gift*, Berkeley: University of California Press.

Tilley, C. (1999), *Metaphor and Material Culture*, Oxford: Blackwell.

Tilley, C. (2000), 'Ethnography and material culture' in P. Atkinson, A. Coffey, S. Delamont, J. and L. Lofland (eds), *Handbook of Ethnography*, London: Sage, 258–72.

The Metaphorical Transformations
of Wala Canoes*

Christopher Tilley

Strathern (1988) has argued that Melanesian people differ from Europeans in understanding relations between producers and products in multiple and divisible terms. She produces an analytical fiction, contrasting an 'ideal type' of a Western body, which is unitary and sexed, with a Melanesian 'ideal' body, which is partible and has both masculine and feminine elements. Bodies become 'male' or 'female' not because of observable biological sexual characteristics but by virtue of the nature of their positioning in social acts. Men and women have both male and female elements which become activated in different social contexts. Gender construction is much more than a simple articulation of difference on pre-existing male or female bodies. It becomes something endowed on persons and their bodies, artefacts, events, architecture and spaces. 'Male' and 'female' are two principles that constitute society; not distinctive attributes of different bodies, but forms of action.

While we may be uneasy with the contrast Strathern draws between Melanesian and Western bodies (cf. Butler 1990) and the associated notion that work or objects cannot be alienated (N. Thomas 1991: 57), what is valuable here, and that which has been cogently stressed in the writings of other Melanesian anthropologists (e.g. Battaglia 1990; Keller 1988; Jolly 1991a; 1991b; MacKenzie 1991; Munn 1977; 1986; Thomas 1995), is a way of thinking about the relationship between producers and their products centring upon *activity*. It is this that produces meanings and serves to gender both persons and artefacts. Objects are created not in contradistinction to persons, but out of persons. A conceptual separation of subject from object (with the latter regarded as inert 'dead' matter) is usefully avoided. Production becomes a performance through which persons and objects create and define each other. The artefact is as crucial and as active a participant in this performance as the person. Production is a practice in which relations with kin, affines and others may be acknowledged, created or marked through a realm of material things. The work *of* men or women may not necessarily be identified as the work of men or women, and the meanings of the artefact may not necessarily be

* C. Tilley, *Metaphor and Material Culture*, Oxford: Blackwell, 1999, pp. 102–32.

grounded in the person of an individual, or in terms of a simple binary opposition between male and female.

In the field of cultural production and reproduction it is the case that an order of artefacts performs its symbolic work of socialization and the creation of social identities silently, continuously and, therefore, relatively unremarkably. Through creating, exchanging and ordering a world of artefacts people create an ordering of the world of social relations. Such a process of objectification is about the construction of meaning and values about social relationships and self-understandings of those meanings and values through material forms. Social images are constructed through the material media of artefacts.

These images and forms of presentation and representation in which persons 'present themselves to themselves' (Strathern 1990: 26) may not be, and often cannot be, articulated in spoken language. The meanings created through artefacts and words cannot be exchanged for each other, and thus the material object forms a powerful metaphorical medium through which people may reflect on their world in a way simply not possible with words alone. Through the artefact, layered and often contradictory sets of meanings can be conveyed simultaneously. The artefact may be inherently ambiguous in its meaning contents precisely because it acts to convey information about a variety of symbolic domains through the same media, and because it may perform the cultural work of revealing fundamental tensions and contradictions in human social experience. In other words the artefact, through its silent 'speech' and 'written' presence, speaks what cannot be spoken, writes what cannot be written, and articulates that which remains conceptually separated in social practice (Tilley 1991). It is a multiple site for the inscription and negotiation of social relations, power and social dynamics. The artefact is both interwoven with and recursively engaged in the production and constitution of society.

In this chapter I consider the metaphoric meanings of the construction of canoes on Wala island, Vanuatu, in the past and in the present, from the general perspective of objectification processes, and as polysemous objects in which notions of place and landscape, rootedness and journeying, exchange, gender and historical meanings become layered and entangled together, drawing on ethnohistorical information and personal field observations.

Canoes as Dominant Symbols

Wala is one of six coral islands, known locally as the Small Islands, situated just off the coast of north-east Malekula, Vanuatu (Fig. 2.1). It is barely 1 km in diameter, fringed by reef and low coral cliffs except for a white sand beach in the south-west facing the mainland of Malekula across a narrow strait. The population of around 200 persons, divided into five exogamous patrilineal clans, is concentrated in one village, Serser, by the beach. People moved to this site, under missionary influence, about sixty years ago. Wala has always been noted as the navigational centre for north-east Malekula, the name 'Wala' meaning 'to run before the wind', 'to sail' or 'to go on an ocean voyage' (Layard 1942: 455). The canoe has always been the largest portable artefact made by Wala islanders and everyone owns or has access to one. Early ethnographers remarked on the sophistication and degree of elaboration of the canoes built on the Small Islands of northern Malekula compared with most of the rest of Vanuatu (Speiser 1990; Haddon 1937). The scene on the beach today with scores of canoes drawn up on the sand is still very similar to that depicted in photographs from the early twentieth century in ethnographic works.

Canoes still remain the principal means by which people reach the mainland, supplemented only very recently by a single small motor boat. Early in the mornings children paddle across the narrow strait to go to the mission school. Catholics paddle over to the church on the mainland for Sunday morning service. People

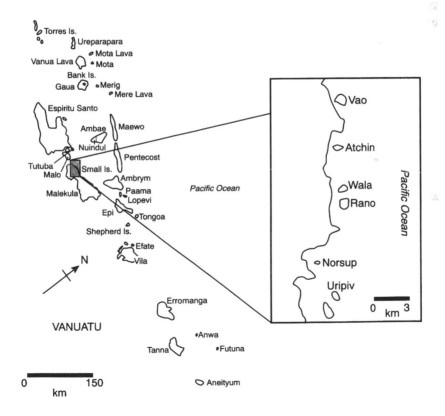

Figure 2.1 The location of Wala and the Small Islands of north-east Malekula, Vanuatu

leave for their gardens on the mainland, returning in the evening with the canoe laden with produce: yams, manioc, coconuts and citrus fruits. Wala has no running water and only limited amounts of rain water are collected. In the past fresh water was brought over to the island by canoe in bamboo shoots, nowadays in plastic containers. Two wells sunk in the village of Serser have water of dubious quality subject to salt-water contamination, and are largely used for washing. Old canoe hulls protect the ridge poles and palm thatch of the houses from the rains. When a person dies the corpse is placed in a canoe during the period of mourning. Paddle dances tell the story of the arrival and departure of canoes. Canoes are used extensively for inshore fishing along the coral reefs and for offshore deep-water fishing. In the past canoes were used to transport smooth stones taken from along river courses on the mainland, to erect along the dancing grounds in the centre of the island. Magic stones, slit drums, pigs, building materials, adzes, indeed most of the resources needed to sustain life and ceremony were, and are, brought to Wala by canoe. With the exception of fishing expeditions, the canoes today are only used for the twenty-minute journey to and from the mainland. In the past they were used in warfare and long voyages were undertaken along the coasts and between the major islands forming the northern part of the Vanuatu archipelago to exchange gifts and acquire tusked pigs. The canoe today has become a living symbol of tradition, of *kastom*.

It is no exaggeration to state that there is hardly any part of life on Wala that is not intimately connected with the canoe. It represents tradition, sustains life, protects dwellings, provides the medium for social contacts, material and spiritual exchange. Precisely because the canoe operates in all these different domains and links them together, it is an artefact invested with considerable symbolic potency: a condenser and transformer of signs, a metaphorical vehicle for transmitting fundamental beliefs and values.

Today, there are three basic types of canoe, all named after the principal tree from which they are constructed (the rav). They all have single outriggers. Three basic forms are distinguished and named after different prow forms:

1. *Rav Msore*: canoes with extended prows with figure-heads.
2. *Rav Solip*: canoes with attached prow figure-heads.
3. *Rav Res*: simple canoes without prow figure-heads.

Virtually all of the approximately 200 canoes which I saw drawn up on the beach on Wala were of the simplest form without prow figure-heads and were undecorated. During my stay on the island I was fortunate to be able to observe six examples of the more complex canoes being built. This was in preparation for the greatest event in the recent history of Wala island: a Sunday afternoon visit by the 'Melanesian Spearhead Group': the ninth annual general meeting of the presidents of the different Melanesian countries. They were to have lunch and conduct a seminar in the small newly established (July 1993) tourist resort on the island, part of a three-day conference being held in northern Malekula. The six traditionally designed and crafted canoes were being built to carry the politicians over to Wala island from the mainland. The canoes had to be 'correct' in every detail and conform to the dictates of *kastom*. The fame of Wala was at stake.

A Floating Forest

In her paper on Gawan canoes of the Massim, Munn (1977) stressed the manner in which fabrication is never simply technological construction, but instead 'developmental symbolic processes that transform both socially significant properties or operational capacities of objects, and significant aspects of the relations between persons and objects' (Munn 1977: 39). This provides the specific starting point for the present analysis. Munn points out that canoe construction, at the most basic symbolic plane, involves taking heavy wood from the bush and converting it into a mobile artefact whose most desirable qualities are lightness and buoyancy. I was told that a canoe should be like a flying fish skimming the ocean waves. Figure 2.2 shows the major named parts of the Wala canoes. Wood and materials from a remarkably wide range of trees and plants are used to construct and decorate

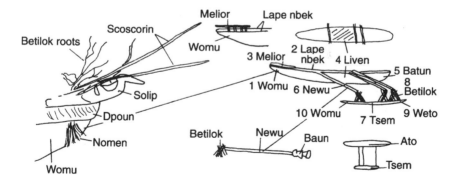

Figure 2.2 Named parts of the Wala canoes (Wala/Rano language). *Ato*: hull. 1: *Womu* (bow). Top of the *rav* ('canoe tree'). A soft whitewood. 5: *Batun* (stern). Base of the trunk of the 'canoe tree'. 7: *Tsem* (outrigger). Young branch of the 'canoe tree'. 2: *Lape Nbek* (beading around the canoe hull). Roots of the 'female' banyan tree. 3: *Melior* (carrying attachment at the bow).Melior tree. Hardwood. 4: *Liven* (central part of hull between outrigger booms). 6: *Newu* (outrigger booms). *Bauer* tree. A hard wood with a red hue. 8: *Betilok* (outrigger attachments). Betilok tree. 9: *Weto* (rear end of outrigger boom). Nomen (tassel made from edible white coconut fibres). *Dpoun* (casing at bow end of canoe from base of coconut leaf where it is attached to the trunk). *Solip* Prow figure-head. Bauer wood. *Scoscorin* (protective branches of the Neor tree over the prow figure-head)

canoes. The hull, – about 4 m long with a maximum width of 45 cm – is made from the hollowed-out trunk of the *rav*, 'canoe tree' (*Gyrocarpus americanus*). The single outrigger – about 2 m in length – is made from a branch of the same tree. It provides a soft whitewood, easily dug out. Today, metal adzes are employed. The traditional technique was to burn out the trunk and cut it with a shell adze. The canoe tree is unmistakable during the dry season, between May and August, when it loses all its leaves. It is largely confined to low-altitude coastal strips up to 150 m. Trees with a suitable trunk shape and size are not found on Wala today and must be taken from the mainland.

The rim of the canoe hull is finished with the addition of a narrow strip of wood taken from the aerial roots of the *nbek* or banyan tree (*Ficus benghalensis*). These utterly distinctive and massive trees, growing to a height of 30 m or more, are the kings and queens of the forest. Aerial roots develop from their branches, descend and take root in the soil below to become new trunks. In this manner the tree can spread laterally almost indefinitely and may, in time, assume the appearance of a dense thicket as a result of the tangle of roots and trunks. They shade all the ancient dancing grounds on Wala (see below).

All canoes have three outrigger booms, about 2 m long and asymmetrically placed, one at the front and two at the back. All three outrigger booms and the attached carved prow figure-head (about 50 cm long) are made from the wood of the *bauer* tree (*Callophyllum inophyllum*). This is a massive seashore tree, growing up to 30 m high, with a crown extending up to 25 m across. Confined to the littoral strip, just above the high-water mark, it has long, spreading branches (the lower ones often horizontal and leaning over the water) and large fleshy leaves. This wood is hard, with a reddish hue.

The prow end of the canoe has two attached carrying handles made from the wood of the *melior* tree (*Dysoxylum gaudichaudianum*). This is a large canopy tree of the lowland forest, resembling a European ash, providing a hard and resilient wood. The outrigger booms are attached to the outrigger by means of strips of wood driven diagonally into holes bored into the float so that the sticks cross one another. The attachment is secured by lashings of coconut rope. The sticks used for this attachment are from the *betilok* (*Myrsine* genus), a birch-like shrub growing in areas of disturbed regenerating forest. The roots of the same tree are attached over the top of the carved prow-head, together with two long wispy branches (the *scoscorin*) from the *neuru* tree (*Casuarina equisetifolia*). This is a tree common to the coastal strip with pendant, needle-like foliage, similar to a conifer and giving it a feathery appearance. It has the ability to grow on the most barren land, such as exposed coral plat forms virtually devoid of soil (Wheatley 1992).

The outrigger booms are lashed to the hull and the outrigger with rope made from coconut fibres. These brown fibres are taken from *long* coconut shells. Other appended tassel decorations at the front and back of the canoe (*nomen*) are taken from the edible fibres of the shell casing of smaller, *rounded* coconuts. The carved prow-head is lashed to the canoe after the wood of the canoe prow has been covered with sheet-like fibre taken from the casing of the coconut leaf at the point where it joins the trunk of the tree.

On special occasions the entire canoe may be beautified with brightly coloured cycas leaves attached to the boom lashings and tucked in around the prow-head. These leaves are widely used as ceremonial body decoration, attached to arm bands and belts.

Thus, no less than eight different plants and trees were used to construct the canoes. There is no functional or utilitarian reason which could explain such a wide choice. Heartwood and sapwood is employed, softwoods and hardwoods, woods of contrasting colour, grain and degree of suppleness, aerial and underground tree roots, young branches and older branches, different fruits from the coconut. The plants and trees employed grow in a wide variety of different environments: in areas of bare coral, in the littoral zone overhanging the sea, on Wala island, on the mainland, away from the beach in the interior of the tropical forest, in areas which are freshly regenerating and in others with long-established growth. The leaves, fruits, bark and other physical characteristics of the trees used (from the deciduous *rav* to the conifer-like *neuru*) are all different. Although people on Wala do not, to my knowledge, make the connection explicitly, the canoe brings together, in one artefact, all the different environmental and physical characteristics of the major trees of the forest, and unites or marries them together as a floating forest. The hull may be beautified with leaves which 'sprout' from it, just as fresh leaves sprout from the canoe tree, at the end of the dry season. The top of the trunk of this tree forms the bow end of the canoe, the base, the stern. Thus, the movement of the canoe on water, bow first, can be regarded as a symbolic transformation of the upward growth of the canoe tree. Both extend themselves in space, sprouting leaves. At one level of symbolism the canoe is a material metaphor for the forest which surrounds and envelops people throughout their daily lives. It is a forest transformed into motion. Houses and villages are rarely visible from the shoreline in Malekula. To an observer standing on the mainland, Wala and the other Small Islands along the coast of Malekula appear uninhabited. From Wala the only visible trace of habitation on Malekula is the large and recent clearing around the Catholic mission and school. The canoe unites the trees of the forest just as it unites people and their life experiences from the moment of birth to the moment of death.

Birds and Roots

The bow of the canoe has a carved figure-head (*solip*) lashed to its tip with coconut rope. The standard Wala island form of this design now consists of two main elements: a bird figure-head with outstretched wings curving backwards embracing a fish. Two canoes being built on the neighbouring island of Rano had variants on this form. One consisted of a bird design only. The other, occurring on the bow of a canoe with an extended prow (*Rav Msore*), consisted of two birds with heads facing away from each other over the top of a pig mounted on another bird. The bird form always occurs and senior men have individual rights to carve variants on this design. The only other design element, carved today, to occur in conjunction with the bird, is the flying fish or the pig. Extending above and out beyond the prow design, over the bird head, two long wispy twigs are affixed and above them again, a tangle of tree roots. The decorated canoe prow-head of hard, red (*bauer* tree) wood contrasts with the creamy yellow-white soft wood of the canoe hull. Traditionally, this figure-head would be blackened by smoking before being attached to the canoe. I saw a number of examples of these blackened bird prow-heads in the houses of senior men, but none were employed on the canoes I saw being built.

The bird design represents the frigate bird (*Fregata ariel*). This bird has a number of fascinating physical and behavioural attributes combining to make it most apposite as the primary, invariant and most significant element in the design of the carved prow. It is a huge sea bird, the largest to occur in Vanuatu, and ranges all over the Pacific Ocean. It keeps mostly out to sea, nesting on small and remote islands. It is mainly black in colour with white patches on the flanks, a great wing span and forked tail. The male has a bright red throat patch which is inflated like a balloon during courtship (Harrison 1983; Bregulla 1992). Frigate birds are superb fliers, soaring magnificently in thermals with little perceptible movement of their wings. They can hover and glide with ease in updrafts along sea cliffs. They have been known to brave cyclones. One of my informants reported seeing them soaring effortlessly over Wala during a violent storm. Another term for the frigate bird is the 'man of war bird' because of its piratical attacks on other fish-eating sea birds. They mob other species, forcing them to disgorge their fish in order to lighten themselves and escape. The fish are swooped up in mid-air by the frigate birds.

The frigate bird thus combines qualities of superb agility, the ability to weather the worst of storms, and voracious habits, terrifying other species. It would be difficult to imagine a more appropriate image for the carved canoe prow-head metaphorically effecting a transference of these desirable qualities to the canoe itself and its crew. The choice of the flying fish design in combination with it reinforces aspects of the same symbolism in relation to the canoe: its desired qualities of buoyancy, being able to glide over the water, and weather storms. The redness of the wood chosen for the prow design strongly indicates that it is the male frigate bird (with a red throat pouch) that is being represented and its aggressive characteristics valorized (see the discussion below).

The frigate bird prow-head is covered with two tentacle-like *neuru* branches and *betilok* roots. The reason for these attachments, I was consistently told, was to protect the carved prow, the most important part of the canoe. Speiser notes that 'it is a gross insult to damage such a prow, and people are said to have been killed for breaking off a boat prow' (Speiser 1990: 225). One informant likened the *neuru* branches to the tentacles of a lobster, protecting it from harm. Such was the almost obsessive concern with the protection of the prow-head that in one case I observed, on the island of Rano, the branches and roots were deemed insufficient and the whole prow end of the canoe had a cage structure of coconut palm branches woven around it which had to be cut loose before the canoe could be launched.

Given that the front of the canoe hull is the part of the tree trunk closest to the sky, the addition of roots in this position, rather than at the stem of the canoe, in effect represents an *inversion* of a literal ordering of nature: roots in the sky, protecting the frigate bird, positioned below. Root imagery carries a heavy symbolic load in north-east Malekula. Roots support, protect and hold up the tree, stimulating growth, making it firm and strong. Their protective powers are clearly being transferred to the canoe and its forward movement. The canoe, unlike the tree, is not anchored to the ground. It is a mobile force which the roots protect as it moves forward. Canoes are always drawn up on the beach with their prows forward, facing the land. The roots are thus highest up on the beach, preventing the canoe from slipping back into the water and being carried away by the waves and the tide. It is interesting to note here that the major reason given for people settling on Wala and the other Small Islands, in the historical past, was the protection this afforded them from attack by inland peoples who were not accustomed to the sea, and who had no means of reaching the islands when the canoes of the inhabitants were all drawn up and 'rooted' on the beach.

In men's traditional club houses on Vao and the other Small Islands of north-east Malekula, as documented by Speiser in 1910, some of the bamboo cane purlins are left with their rootstocks intact, projecting beyond the front edge of the roof (Speiser 1990: 105). Ancestor houses lining the dancing grounds also had a ridgepole consisting of a trunk, complete with roots. The root end pointed forward toward the interior of the dancing grounds. The roots on both the club houses and the ancestor houses represent the opened wings of a bird. The body of the bird is formed by the ridgepole itself (Speiser 1990: 348). Root imagery is thus connected both with height and birds. It links canoe prows, the highest part of the canoe, the gable ends of men's club houses, and the ridgepoles of ancestor houses. The ridges of the club houses were themselves covered by the hulls of old canoes (Speiser 1990: 105), as houses are today. And all these are built exclusively by men. The ridgepoles, with their bird-roots, hold up and protect the men's club house and houses containing the ancestors and ancestral powers, just as the canoe protects those using them. In turn, the canoe covers and protects the ridge of the club house. Height is connected with spiritual power, a dominant theme in Wala culture. The bird depicted by the rootstocks of the houses is the soaring hawk. So the roots over the canoe prow may also be interpreted as representing a hawk soaring over the frigate bird. The hawk is the most ritually important and symbolically significant bird connected with the land. Just like the frigate bird it has ferocious habits, killing other birds. The frigate bird and hawk symbolism of the canoe prow-head thus connects the domains of land and sea and links them to the communal club house in which men sit together, as if in a canoe. The prow adornment is a metaphorical vehicle of meaning that synthesizes, through metonymic association, non-human and human domains, rootedness and movement, land and sea.

On Tanna island in the south of Vanuatu, Bonnemaison (1994) has shown that the canoe is both a metaphor for society and sociality: 'the individuals who meet daily make up the core of a local group, a "canoe" or *niko* as it is called on the island, which itself is linked to a larger "canoe" made up of several patrilocal clans sharing the same territory and mythical heritage' (Bonnemaison 1994: 108). Munn has argued in relation to the island of Gawa, Milne Bay province, Papua New Guinea, that a basic contrast exists between the body-house and the body-sea. These represent opposite poles along an axis of the spatio-temporal extension of the self. The body-sea, in the form of the Gawan canoe, is an exterior self-extending dynamic pole of social being because the canoe establishes social connectivities. The house, on the other hand, is an interiorized bounded world, statically rooted in the heaviness of the ground (Munn 1986: 79).

On Wala the canoe, the club house and its society of men appear to be fundamental metaphorical transformations of each other. The canoe is a floating communal men's house with its roof turned upside down on the water. The club house is conversely an inverted canoe which can no longer float. The spatio-temporal transformation of the self and the move from the interior social world of the men's club house outwards to establish relationships with persons on other islands requires that the club house be 'opened out' with the men now sitting in its roof.

A basic contrast between rootedness and journeying is at the heart of much Melanesian thought (Munn 1977, Battaglia 1990, Bonnemaison 1994: 122). Ultimately, on Vanuatu, the strength of custom or tradition is symbolized by the firmly rooted banyan tree (Bonnemaison 1994: 122, Jolly 1982). Personal identities are rooted in the land. The person who abides and stands straight will take root just like the tree. Conflicts occur because people leave their roots. A place provides persons with roots, the canoe forges alliances and relationships necessary for social and spiritual reproduction. Thus, on one symbolic plane the prow imagery of the Wala canoes unites the *duality* of the construction of social identities which are both permanently rooted in the land and established through movement over the sea. The land is the terrain over which the hawk soars and the ocean is where the frigate bird has its domain.

Figure 2.3 Wala canoe prow figure-head

Canoes and the Body

Below the carved prow design, and attached through holes bored into the canoe hull, two fan-like sets of coconut-fibre decorations are suspended. These are also suspended from holes bored in the bow end of the canoe hull, which has no other decoration. On three of the six canoes I observed the coconut-fibre decorations used at the prow end of the canoe were white in colour, contrasting with the red-brown of the coconut rope lashings. Those at the bow of the canoe were the same, colour as the rope lashings (Fig. 2.3). These two types of coconut-fibre decorations, with white/red colour contrast, are derived from different coconuts. Those attached to the bow end of the canoe are derived from small, round, sweet coconuts whose fibres are edible. Those at the hull are derived from long coconuts whose inedible fibres are also used for the canoe lashings and bindings. The white fibres at the front are white and old, those at the back fresh and young. The coconut fibres dry out from the centre of the coconut. Those used at the front come from the inside of the husk, next to the kernel, those at the back from the outside. So there are a series of structured contrasts at work here:

white	red
old	young
dry	wet

inside	outside
edible	inedible
short	long
CANOE BOW	CANOE STERN

The tassels at both the bow and the stern were consistently described to me as being the 'moustache of the canoe'. The term 'moustache' may thus refer to the canoe as being both double-headed and gendered as male. Layard records that large sea-going canoes which were constructed formerly (see below) had bows at each end and two carved prow-heads. In addition, these canoes had two subsidiary figure-heads, one on each side of the main figure-head (Layard 1942: 460). It is possible that the moustache tassels on either side of the bow and stern and below the prow figure-head are transformations of this arrangement on the large sea-going canoes.

When I asked why the fibres were white at the bow and red at the stern of the hull, I was told by the group of men making the canoes that this would make the bow end appear 'flashy' and 'stylish'. I asked whether the presence of the 'moustache' at both ends meant the canoe had two heads or faces, and meant that the canoe was a double-headed male. This suggestion caused much amusement. The explanation given for the canoe 'moustache' was that it was a traditional *kastom* element. On long sea journeys the moustache would be used for lighting fires when people reached the land. It became quite apparent that such an explanation could not be given much credence. In motion both canoe 'moustaches' were continuously drenched by even the slightest of waves and would hardly provide effective tinder. Apart from the reference to *kastom* and the 'stylishness' of the moustaches no other explanations were offered to me by the group of men building the canoes.

One informant living on the mainland opposite Wala gave me an altogether different explanation. The canoe represented a big man. The moustache at the front was the ears of the big man, who listens to others, that at the back his *namba* or penis sheath. The outrigger booms were his arms and legs, their attachments to the outrigger his fingers and hands. The outrigger itself represented the sole of the foot and the palm of the hand. The bird prow beak symbolized the mouth of the big man, his skill as an orator. Finally, the banyan root lashed around the rim of the canoe hull was the belt of the big man holding his *namba* at the stern end in place (Fig. 2.4). Why, I asked, if this was the case, did the canoe only have three rather than four outrigger booms, two at the back (two legs) and one at the front (one arm)? And why was the big man's mouth (bird beak) closed? Two explanations were provided: large sea-going canoes, built in the past, would have four outrigger booms. This was not possible in the case of the smaller coastal canoes being constructed today, as four booms would interfere with paddling. Bird prow designs, which could only be carved by senior men, who jealously guarded these

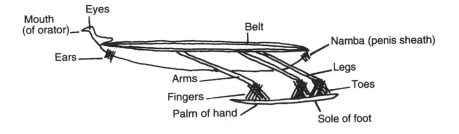

Figure 2.4 The canoe as a big man

rights, were double, with the beak opening extending down the gullet. I had previously been shown these double bird prow-heads in the house of one of the five Wala 'chiefs' who owned the rights to carve it. Speiser (1990) also provides an illustration of one example from eastern Malekula, collected *c*.1910 (Fig. 2.5). Layard (1942: 456) states that the larger sea-going canoes constructed in the past had four booms (see below). This interpretation of the canoe as a dressed male body makes good sense of the two 'moustaches' and their different colours. It also provides an explanation of the contrasts noted above: the 'edibility' of the moustache (ears) hanging beneath the prow (in the historical context of a culture with a strong element of cannibalism) and the 'inedibility' of the moustache (penis wrapping) at the bow. The white fibres at the prow end may be further understood as signifying the age of the big man (white facial hair). The 'ears' at the bow end are shorter than the *namba* which was traditionally dyed red (Layard 1942: 42), with wood frequently preferred to bast for the belt (Speiser 1990: 175). Today, *nambas*, red or purple in colour, are still worn on ceremonial occasions and in dance displays for tourists. Cycas leaves are tucked into the belt and under arm rings and leg rings. These same leaves were stuck under the banyan wood 'belt' of the canoe to beautify it during the journey taking the Melanesian heads of state to Wala island. The canoe, in this interpretation, is a heavily anthropomorphized form, the body of a big man. It directly projects an image of the big man adorned. For ceremonial occasions, both the human body and the body of the canoe are decorated to create the best possible impression on guests. Such impression management is essential to create the desired effect: the seduction of persons with whom one wishes to maintain links and from whom one wishes to extract gifts and favours.

When the Melanesian heads of state had landed on the island, they were led from the beach into the dining hall of the Wala resort by a group of male dancers dressed in *nambas* with leaf decorations in arm and leg rings. They performed a traditional welcome dance: the paddle dance. One man at the front held a canoe prow-head and moved forward away from the beach. Two parallel rows of male

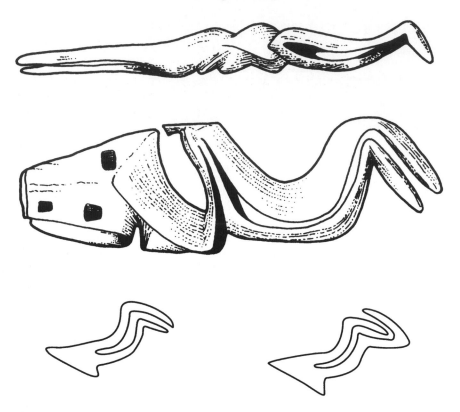

Figure 2.5 Top: traditional carved prow-head collected from Vao island (*c*.1910). *Source*: Speiser 1990, plate 64. Bottom: sketches of elaborated bird figure-heads from Atchin island (*c*.1914). *Source*: Layard n.d.

dancers followed, holding brightly decorated paddles representing each side of the canoe. A man at the rear symbolized the canoe rudder. Having arrived by canoe the heads of state were 'carried' by the canoe of men into the dining hall of the tourist resort. This was built in the form of a men's club house (i.e. an inverted canoe). Afterwards the paddle dancers carried them back to the water. While they had set foot on land, they had never left a canoe. The canoe, as decorated artefact, metamorphosed into the canoe as a decorated body of men, into the canoe of the dining hall, and back again.

An Androgynous Object

From the discussion so far the Wala canoe would appear to be an artefact entirely gendered as male. Its exclusive construction by men, the vicious and violent frigate bird (and hawk?) imagery, references to moustaches and *nambas*, would all seem to indicate that the canoe is masculine labour objectified, acting as an icon of male power and virility. Yet there appears to be another less obvious and more subtle level of meaning, quite literally *embedded* in the choice of wood and canoe shape. One informant told me that the best wood to use to construct the hull would come from the female *rav* tree at the time that it fruited. Such a canoe would be stronger, last longer and not be so heavy. The outrigger should be made from the younger branches of the tree. The hull and the outrigger, then, connote imagery of lightness, fertility and femaleness. A cultural distinction is made by Wala islanders between male and female banyan trees. The aerial roots of the male banyan are said to grow down parallel with the trunk, those of the female to grow outwards so as to create cave or womb-like hollows. It is wood from the *female* banyan root that is wrapped around the canoe hull. While roots representing the hawk protect the frigate bird prow-head (both embodiments of an aggressive masculinity), it is the elastic female banyan root that is wrapped around the hull, protecting the people contained within it. The canoe, long, narrow and vulva-shaped, contains people like a womb containing the child. At a more abstract level a female principle holds up both men and women. It is this that is fundamental, while the overt male imagery of the prow-head is merely an appended form. Traditionally, if a man and a woman travel together in a coastal canoe the woman should sit at the back, at the stern (stump/root) end of the hull, and steer. The man at the front may propel the canoe forward, and be most visible, but the woman *governs* its course.

MacKenzie (1991) has discussed in detail the manner in which string bags constitute a subtle medium for Telefol people of the New Guinea Highlands to articulate views and images of relations between men and women. She argues that the *bilum* is used in the construction of identity to *both* reinforce and blur, or neutralize, an obvious oppositional separation between men and women. It points '*beyond* the polarities associated with the contrasted concerns of the sexes to an identity of interest between them' (MacKenzie: 193). The androgyny of the Telefol string bag becomes a material metaphor for reflecting on society and the cosmos.

Something similar seems evident in the case of the Wala canoe, but the female elements are far less obvious and far more understated compared with the overt and brash male prow imagery. The female elements in the canoe constitute a silent unremarked complementary material discourse which nevertheless remains fundamental to the ontological significance and value of the artefact. Although the canoe itself is exclusively fabricated by men, its manufacture is continuously nurtured by women who cook on a daily basis for those involved in its construction. In the past women also plaited the sails which propelled the large ocean-going canoes forward (see below).

Ethnohistory of the Wala Canoe

In the past two kinds of canoes were constructed on the Small Islands of Malekula:
(1) the smaller coastal canoe, still used and constructed today and (2) large sea-
going canoes used for exchange and ceremonial expeditions. These were already
being replaced by whalers purchased from white traders in the first two decades
of the twentieth century. Speiser's (*c.*1910) and Layard's (*c.*1914) descriptions of
the smaller coastal canoes of north-east Malekula match well the manner in which
they are constructed today in terms of size, number of outrigger booms and bird
prow-head, for example (Speiser 1990: 225; Layard 1942: 458; Layard n.d.). The
coastal canoe, single-ended and with a single figure-head on the bow, could sail
in one direction only, whereas the sea-going canoes, double-ended with main and
subsidiary figure-heads at either end were constructed so as to be able to sail in
both directions. While the coastal canoe had three booms, the two aft booms being
close together and at some distance from the fore-boom, the sea-going canoe had
four equidistant booms close together amidships. This symmetrical arrangement
(see Fig. 2.6) was clearly required by its reversible direction. The large sea-going
canoes usually had two wash strakes while the coastal canoe was usually a simple

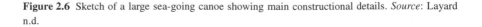

Figure 2.6 Sketch of a large sea-going canoe showing main constructional details. *Source*: Layard
n.d.

dug-out with a rail along each gunwale to which the booms of the sails were lashed. Layard states with regard to the coastal canoes: 'I have no special notes regarding the construction of these small craft, nor of the rites accompanying it, save that, trees large enough being found on all the Small Islands, all the work is done at home . . . the work on the canoe is a purely family affair (Layard n.d.: 67–9). Speiser's and Layard's descriptions of large sea-going canoes from Vao island do not mention the presence of roots above the prow-head, or 'moustaches' at both ends of the hull, or the different types of wood used in canoe construction. Layard's unpublished manuscript (Layard n.d.) about canoe construction on Atchin island (a short distance to the north-east of Wala) mentions that 'sticks of the *nilak* plant pulled up by the roots and attached root-forward are made fast, for magical purposes, alongside the subsidiary figure-head' (Layard n.d.: 4). I have already noted that it is possible that the 'moustaches' of the contemporary canoes represent the subsidiary figure-heads present on the large ocean-going vessels of the past, but since all the other details of the canoes conform to Speiser's and Layard's descriptions this seems unlikely. Layard records that the small coastal canoes from Atchin island had bird figure-heads divided into two main classes. One was a plain figure-head with an ordinary beak ending at the throat. These could be made and used by any man without payment and, consequently, were rare since virtually every man desired something more elaborate. These are the usual form carved today (see Figs. 2.2 and 2.3). The second was a bird figure-head of the same form but with the split forming the beak extending right down the neck. The right to carve this design had to be purchased from someone already possessing it (see Fig. 2.5). Additional features such as a pig or pig's snout, or rarely, a human face might be carved behind the neck, each entailing payment to a previous owner. Another form Layard describes as a 're-entrant-tusker figure-head', of the same form as the elaborated figure-head, with extended beak, but with the addition that the upper portion of the beak itself, at its extremity, is extended backwards over and above the upper portion (Layard n.d.: 72; see also Fig. 2.5). There are clear parallels here between the degree of beak and neck elaboration of the bird figure-heads and the practice of producing tusked boars by knocking out the upper incisor and allowing the lower to grow unhindered, spiralling through the jaw (see below). Bird-beak and boar-tusk imagery become conjoined. Neither Speiser nor Layard record the bird species represented by the figure-head. The contemporary presence of a flying fish behind the plain frigate bird prow figure-head, carved today without an extended throat slit, seems to be a recent innovation, and is related to the carving of prow-heads as pieces of tourist art: a single bird design not being considered sufficiently elaborate to satisfy this market.

Speiser mentions three features not present on the small canoes observable today on the Small Islands, and those examples I watched being constructed on Wala. The first is the double bird prow-head. The second is the presence of small cross-boards fixed 'on the bow-like little seats. They have nothing to do with navigation but simply indicate the *suque* [*maki*] rank of the owner. The higher his rank, the larger the number of boards he may display on his boat' (Speiser 1990: 225). The third is the occurrence of sails on the small coastal canoes. These were made from the leaf sheaths of the coconut palm sewn together with fibres forming a triangular surface stretched between two crossed bamboo poles. The upper edge was curved.

> Each bamboo is stayed by three coconut ropes. From one bamboo a rope runs to the rear outrigger spar, the second to the front one, and the third to the bow. . . . If the sail is not being used, it is rolled up and tied with broad tapes of bast. When the sail is in use, these tapes then hang down fluttering from the bamboos and are an additional adornment to the boat. (Speiser 1990: 225)

Layard records that the Vao word for sail means 'wing' and each upper angle of the sail was referred to as the 'tip of the wing' (Layard 1942: 461). From these observations it would appear that the canoe sail represented a large bird with outstretched wing tips, either a frigate bird or a hawk. The redness of the sail might indicate the former.

These canoes could not tack against the wind because of their shallow draught, but they could move at an extraordinary speed with the wind. As is the case today, they were owned by individual men or families and were not used as items of exchange. Layard notes that some large coastal canoes were heightened by planks lashed to either side of the hull with a special bow piece added in the form of a short piece of hollowed log (Layard 1942: 458). A hollowed log was used to heighten the bow end of two of the canoes (*Rav Msore*) that I saw being constructed, but planks were not added along the sides of the hull.

Canoes today are only intended for short trips between Wala and the mainland. The much larger sea-going canoes constructed in the past were used for voyages between Wala and the other large islands in central and northern Vanuatu: Ambae, Espiritu Santo, Pentecost, Ambrym, Maewo and Malo. The distances involved were up to 80 km, generally involving island hopping. The scale and duration of these voyages always seems to have been rather limited compared with other areas of Melanesia and Polynesia, with long-distance voyages not being undertaken, and rarely by night. From the 1870s onwards the frequency of these contacts was progressively reduced as a result of frequent blackbirding (canoes provided an easy target for the slave labour ships) and the gradual introduction of inter-island steamers subsequently.

These large canoes could hold up to forty persons. Somerville (1894: 375) reports that they were built of three or four planks each about 30cm wide, lashed together with tough fibre, caulked with tree gum and painted with lime. The bow ends were formed by the keel foundation log, sharply pointed off. The planks did not overlap and were sewn at their ends to a thick heavy end board extending to 30 cm or so above the planking, narrower at the base and wider at the top, 'forming a sort of trough with spreading sides'. These canoes had a mast:

> it is generally a stout, moderately straight piece of bough with a fork or 'yaws' (as on a gaff or boom) at the foot which rests on a stout transverse stick like a thwart, so that the mast does not touch the bottom of the canoe. It is supported on this thwart by the fibre guys fore and aft, and the sail is hoisted on it, the halliards reeving through a hole burnt in the head. By this means, the mast can be 'stepped' at any part of the boat, there being several of these thwarts, acting as strengtheners, and can be inclined at any angle towards bow or stern, and lowered quickly in case of a squall. (Somersville 1894: 375)

The large sea-going canoes were used exclusively by men and were both communally constructed and owned. For Vao, Layard notes:

> if a *maki* is in progress, the canoe belongs exclusively to the officiating 'line' of Makimen, that is to say, to the officiating matrilineal moiety. . . . Though the canoe belongs to the whole 'line' . . . one man is spoken of as the 'owner.' It is he who performs the main sacrifices and directs operations, and it is with his personality that, after the consecration rite, the canoe itself becomes identified. . . . This man is probably the leader of the Maki. (Layard 1942: 462–3)

Women could not touch or enter these canoes. Codrington (1881: 293) records that the large canoes had proper names, like a person. By the turn of the century these were replaced by European whaleboats, bought from white traders. On Vao the first such boat was bought in 1906 and the big outrigger canoes left to rot on the beach (Speiser 1990: 223). The main purpose of these voyages to the other large islands of northern Vanuatu was to acquire tusked boars for the great *Maki* (grade-taking ceremonies). Another essential item, turmeric (used to dye clothing), was obtained exclusively from the island of Malo and southern Espiritu Santo. Because of frequent fighting they were also known as war canoes (Layard 1942: 456). As regards the sails, Somerville comments that 'all down both of the outside edges of the sail long graceful fringes hang down, with extra large bunches at the tips' (Somerville 1894: 376). Layard records that the sails were of plaited pandanus leaves, dyed red. The plaiting was done by the women, and the sail-strips sewn together by the men (Layard 1942: 460). Layard also records that on Atchin island the task of sewing the strips together was performed by the men in the dancing ground in secrecy (Layard n.d.: 35), transforming an undifferentiated female labour product into the finished form of the sail, which was attached to booms and displayed in the dancing ground before being carried down to the beach and

erected on the canoe. These accounts provide further evidence for the comple-
mentarity of male and female labour in canoe construction and confirmation that
the sail symbolized a large bird's wing, the fringes representing the frayed plumage
of the primary feathers on the wing tips.

Elaborate rites and magic accompanied various stages of canoe construction
from the initial tree felling to its consecration when completed. These rites were
most elaborate for the large sea-going canoes. They varied in their details between
each of the Small Islands of north-east Malekula, and from one village to another.
The general pattern and purpose appears, however, to have been similar: (1) to
provide payments to those involved in the work and to the owner of the land on
which the tree was felled; (2) progressively to transform an *object into a subject*.
As the canoe was successively constructed the rites accompanying the process
became increasingly more elaborate, culminating in the consecration rites
accompanying its launching.

Proceedings opened with a feast in the dancing ground. A small initiatory sacrifice, usually of a fowl, was performed at the foot of the tree to be felled, almost
invariably on the Malekulan mainland. After the tree felling, pigs were killed and
presented along with yams to the workers.

The log was then dug out and payments of pigs, including tuskers, made to
indemnify the owner of the land on which the tree grew. The dug-out was then
transported to the island and more pigs were sacrificed (but no high-ranking
tuskers). Bow and stern pieces were cut and fitted but not made fast, accompanied
by more feasting. The exact placing of the outrigger booms on the hull required a
ritual specialist, who worked in secret, cast spells and placed the booms where they
were to be attached. The rest of the canoe was finished. After completion another
rite followed involving the ritual specialist placing a sow in the canoe and forcing
it to run up and down from one end to another, symbolizing the fertility of pigs
and auguring success in pig exchanges. The sail was then assembled, followed by
an all-night inauguration dance ('the dance of the sail') in the dancing ground.
More pigs were sacrificed at either end of the canoe below and around the figure-
heads, and a taboo put on women entering the canoe (Layard 1942: 464ff.).
Accompanying the entire process of canoe construction, presentations of cooked
food, provided by women, were made to the workers and divided between the
different village segments involved in canoe construction. Godefroy records that
for the consecration of a whaler in 1926 the rites performed were 'similar to that
of one of the great chiefs' (i.e. men of high *Maki* rank) (Godefroy, cited in Layard
1942: 468). Separate rites were performed to consecrate the hull, the sail, the mast
and the rudder. Through these rites, in particular pig sacrifice, the canoe acquired
a *subjective* identity. It was given a proper name, like a person. Through the sacrifice of boars it acquired a high rank, like a big man and a 'soul', and was gendered
as male.

It is interesting to note that the initial sacrifice at the base of the tree to be felled was of a fowl, the sacrificial bird used for female grade-taking rituals. Once the tree had been felled, pigs and tusked boars, quintessentially associated with men and male grade-taking ceremonies, were used: female substance was converted into male essence. Special houses were built on the beach to protect the canoes and, subsequently, the whalers. These vessels were considered not only to live like human beings of high rank (in a large house), but also to die, and mortuary rites appropriate to the status of a high-ranking big man were performed for *wrecked* vessels. Those which were not wrecked during their life time were allowed to die a 'natural' death, i.e. to slowly rot away in their tabooed boat house. Timber was never taken away for firewood or any other use (Layard 1942: 470–2). One of my informants remarked that it was 'taboo to sleep in a canoe'. From the above account the reason for this seems faintly evident. The entire process of canoe construction converts inanimate materials into an embodied subject. In the process fixed trees, rooted in the ground, are converted into a mobile subject. Thus, it is only possible to sleep in a canoe when dead.

Model mortuary canoes in the Small Islands were used to finish the memory of the dead. As a final act in mortuary ceremonies, the body of the dead was metamorphosed into a gift of food to maternal affines. A model canoe, loaded down with a pig, yams, bananas, coconuts, puddings and fowl, is sent away across the water to maternal relatives on a neighbouring island or on the mainland as the final act in mortuary ceremonies. One of my informants remembered having picked up such a canoe, sent over to Wala from Rano island in 1977. The food was used by the maternal relatives, the model canoe itself left to rot on a small, rough-stone platform on the beach.

Dancing in Csanoes

In the centre of Wala island there are five great *namel* (dancing grounds): They are all built high up on the island, out of sight of the sea. Long parallel lines of megalithic structures consisting of rows of stone tables and monoliths line the dancing grounds, the rows of monoliths representing genealogical depth: successive generations who have taken part in the Maki or grade-taking ceremonies in which individuals acquired new names and statuses, involving the sacrifice of tusked boars (for men) and fowls (for women) to ancestral spirits said to enter the stones once sacrifice had been performed. Each *namel* was associated with a communal *nakamal* (men's club house). Enormous banyan trees still shade these places. Each of the dancing grounds belonged to a neighbouring village. One side of the dancing ground, the upper side, marked by large coral and stone monoliths (up to 2 m high) was the men's, the other marked by much smaller monoliths, the lower side, was the women's. These places were the centres not only for the Maki rites (which have not taken place in living memory), but also for the entire ritual life of the village community and were of paramount significance.

What is of great interest here is that the dancing grounds on Wala are all shaped like canoes. Figure 2.7 shows a sketch plan of Lowo *namel*. It is long (about 140 m) and linear in form, and widest (about 24 m) in the centre. At the western end this dancing ground even has a prow-like structure, diverging to the north of the main line of stones on the men's side at an angle of about 45 degrees, consisting of two parallel lines of monoliths and stone tables and terminating with a single large monolith of smooth stone taken from a river channel on the mainland.

Layard does not mention the overall canoe-like morphology of the dancing grounds with reference to Vao island (and here they may very well be a different shape), but notes other connections between their spatial organization and canoes. In the great *Maki* ceremonies the arrangement of stone monuments and tuskers was likened by his informants to a large sea-going canoe. Circle tuskers were added to represent bird figure-heads at either end of the hull. Tusked boars on the upper or men's side of the dancing ground were likened to the body of the canoe, while gelded tuskers on the lower or women's side represent the outrigger float. Special high-grade tusked pigs on stone platforms behind the stones on the upper side were likened to piles of food, principally yams and bananas, placed on the lee platform of the canoe (Layard 1942: 428) (Fig. 2.8). Symbolic weight is thus placed on the male side of the dancing ground, laden down with high-grade tusked pigs and yams. In this manner the canoe hull and prow-heads become identified

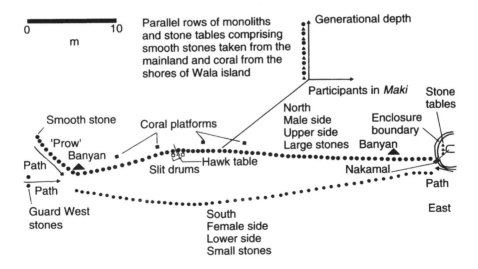

Figure 2.7 Sketch plan of Lowo *namel* (dancing ground), Wala island

as male, the outrigger as female. The androgynous nature of the canoe as an indissoluble mixture of male and female elements becomes even more apparent through this symbolism. It is also possible to understand why a ritual specialist was required secretly to bespell and precisely position the outrigger booms on the hull, because it was not just 'arms' and 'legs' that were being attached but 'male' to 'female' in complementary opposition.

The culminating parts of the Maki grade-taking ceremonies were likened to the approach of a canoe, and Layard notes a general similarity in the ceremonial pattern for the *Maki* rites, canoe inauguration and gong raising. Layard suggests that in the past a new canoe might have been constructed with each successive *Maki* (Layard n.d.: 16). The chief use of the canoes was to obtain pigs and the chief use of the pigs was in Maki sacrifice. Pig sacrifice was essential to all the main stages in the production of the canoe. Pigs made canoes just as they made men's souls.

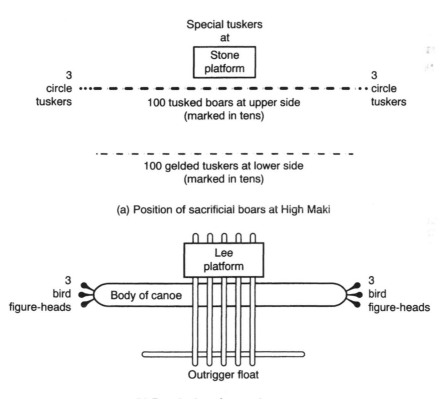

Figure 2.8 Diagram illustrating the comparison made by Layard's informants (*c.*1914) between the arrangement of sacrificial pigs at the *Maki* ceremonies and a large sea-going canoe. *Source*: Layard 1942, Figure 52

An entire *Maki* rite would cover about thirty years, or the length of a generation, and the life of a canoe would have been about the same.

Another feature of the dancing grounds is of interest in this connection. At Lowo and at the other *namel* on Wala, it is the 'female' banyan tree that dominates. These huge trees shade and protect the dancing grounds under which the grand ceremonies were performed. The enormous root systems of these trees wrap themselves around and incorporate the ancestral stones. In an analogous manner the roots of the female banyan are wrapped around the hulls of the contemporary Wala canoes. While Wala islanders make a cultural distinction between the male and female banyan, in practice it is rather difficult, especially for an outsider, to tell the two apart. More to the point, the 'gender' of these trees is highly ambiguous. Through time, with the continued growth of aerial roots, a 'male' banyan with roots growing parallel to the trunk, will transform into a 'female' banyan: its gender is both inherently mobile and open to various interpretations.

Slit Drums and Canoes

In the centre of the men's side of the dancing ground huge slit drums were erected, some twice as tall as a person. The rites accompanying the erection of these drums were 'just like those for gong raising' (informant cited in Layard 1942: 464). The same hardwood trees were traditionally used to make both the large sea-going canoes and the slit drums, and both were dug out in similar ways with shell adzes. Like the canoes, the drums are considered the exclusive property of men. These drums, some of which have now been freshly erected in the centre of three of the dancing grounds on Wala (of the 'wrong' kind of wood – soft whitewood of the same tree used to construct the coastal canoes today), are slit longitudinally with solid sections at the top and bottom, with a face carved at the top. Traditionally, the slit forms the mouth of the face, with lips on either side. The sound which the drums emit is said to represent the voice of the ancestors.

Clausen interprets these drums as 'hermaphrodite':

> the slit now held in the patrilineal small islands to be a mouth with its two lips may originally have represented rather a vagina with its two lips. Such an interpretation would, in fact, be more consonant with the vertical position of the slit which extends right down near the base of the drum representing the human figure than with its interpretation as a mouth which physiologically should be at right angles and not vertical to the body of the instrument. (Clausen 1960: 18)

It seems clear that the slit drums, like the contemporary canoes and the dancing grounds, are *androgynous* forms: both penis and womb-like, with male 'lips' and a female 'vulva' combined into one. And, like the canoes, they are conceived as subjects rather than objects, their voices called to speak by the drummer.

The intimate connection between these drums and the canoes in terms of mode of construction, type of tree used, rites performed, and their androgynous nature is striking. There is another connection. The canoes transport people over the ocean. The slit drums are

> like canoes on the ocean of the spirit world, connecting mankind with the otherwise unknown world of the ancestral ghosts whose voices they represent and who live on – literally, feed on – the psychic essence of the boars that are sacrificed to them. Slit-drums thus constitute a psychic medium of communication with the spirit world corresponding to the trading voyages of the canoes in external life. (Clausen 1960: 19)

Clausen, following Layard (both he and Layard regarded the canoes as embodying exclusively male symbolism), interprets the 'female component' in the slit drums in terms of an evolutionary developmental sequence on the Small Islands from an originary matrilineal social organization to a patrilineal and patrilocal one, which already existed when Layard carried out his research in 1913–14. The overt male ideology of the canoes and the slit drums was all to do with men freeing themselves from women and their powers and asserting their own dominance. The 'hermaphroditic' nature of the slit drums was thus a kind of relic of a previous social order. Abandoning this evolutionary framework, another potentially more insightful interpretation is that the female connotations of the slit drums, and those identified above in the contemporary canoes, are part of a discourse about both male and female essences or powers, which are being brought together and combined in the silent discourse of material culture, a discourse which is not and cannot be articulated in language but is nevertheless made continuously present through the objects themselves. The canoes, the dancing grounds and the slit drums are ideal objects for this purpose, since none could be, or were, exchanged. They constituted permanent markers of the inalienable wealth (Weiner 1992) of Wala islanders.

Conclusion: From Tuskers to Tourists

I have argued that contemporary canoe building involves the material surfacing and articulation of a series of material metaphors bound up with the creation of social identities and intertwined male and female essences. A hierarchy of male power is implied by the overt canoe imagery, the distinction between the upper and lower sides of the dancing grounds and in the slit drums. Yet at a base metaphorical level the canoe hull and outrigger are female, the female banyan tree protects the dancing grounds, its root is wrapped around the canoe hull, ancestral voices issue from the vaginal mouth of the slit drums. Wala canoe imagery, past and present, combines both male and female metaphorical imagery, male and female work. They are ambiguous artefacts in which various levels of meaning appear to contradict each other: the overt male imagery of the canoe is quite literally wrapped around with female forms.

Among the traditional beliefs of Wala islanders was the notion that the human body was made up of a hard, dry, male component (the bones of the skeleton), wrapped around by wet female flesh and blood. Layard notes that in this body image symbolism

> in which the body symbolises the whole psyche, the matrilineal line of flesh and blood symbolises, or is the carrier of, the soul which is female. The bones and patrilineal line of descent symbolise the spirit which is not only incubated in the soul, but, like a skeleton, upholds it and gives it form, and subsequently out lives the soul, which, having performed its incubating function, dies, giving way to the male spirit which is at once its father and its child. (Layard 1972: 321)

The combined male and female imagery of the canoe may thus ultimately be a material metaphor for marital bonds and procreation, with the canoe representing the reproduction of society as much as an individualized body-subject.

The men doing the work of constructing contemporary Wala canoes often preferred to give practical and 'functional' explanations for even such obviously symbolic items as the canoe 'moustaches'. A reluctance to talk about the meanings of material forms has both been noted by, and frustrated, many Melanesian anthropologists (e.g. Forge 1979; Keesing 1987). This refusal to verbalize is perhaps because the artefact as material metaphor does the talking in a much more profound, succinct and vivid manner. The canoe is 'good to think' precisely because its meanings are not verbalized. Discourse occurs at the silent level of the artefact and is continuously presenced in the world as such. It is a discourse which is not, and cannot be, articulated in speech, in a social world constituted, as elsewhere in Melanesia, by extreme male/female sexual antagonism. The symbolic work the canoe does is to 'resolve' or contain, within the image of the largest and arguably the most important portable artefact, contradictions in social life, relations between men and women, which cannot be discussed in language or negotiated in social practice.

Today, none of the traditional rites, described above, linked with the construction and consecration of the large sea-going canoes, are performed. I was present at the launch of one of the new canoes and was invited to sit in the front. The canoe was pushed off the beach, with one of the five Wala chiefs in the rear steering. We followed the coast of Wala for about ten minutes and then returned to the launch site, dragged the canoe up on the beach, and then left to drink kava. It would be easy to describe the coastal canoes constructed on Wala island today as empty vessels of custom. They have been decontextualized from the ritual practices surrounding their construction and consecration. Mortuary ceremonies are no longer performed for them. None of the canoes on Wala island, except for the six

I watched being built, even had carved prow-heads, although these were being carved as tourist art in limited numbers for a growing tourist market.

The reason for these canoes being constructed at all was the visit of the Melanesian heads of state to the island. Subsequently, they were to be used to ferry tourists over to the new resort. The knowledge of the meanings of the details of the canoe design has an entangled history. Coastal canoe construction on Wala island has been continuous (whitewood canoes only last for four to five years). We are not simply dealing here with a reinvented tradition of canoe manufacture, but the metaphorical meanings of the artefact have clearly not remained constant, nor are they consistent. They invite reinterpretation by Wala islanders and by myself.

The enduring symbolic and social significance of the canoe for Wala islanders has always principally resided in its use as a vehicle of power, and in the social *relationships* that it engenders. The process of constructing canoes today brings forth and acts so as to negotiate social relationships between men. Those building the Wala canoes (up to ten being involved at anyone time) belonged to all five patriclans on the island, among whom in the past warfare took place, and between whom in the present, disputes over land and resources are a recurrent element. The project of building the canoes was bound up with a perceived need to both preserve and revive *kastom*. It forged a remarkable sense of solidarity and community spirit between the men engaged in the act of construction. It is of interest to note here that the old traditional *personal* insignia of rank, such as special carved prow-heads or wooden boards lain across the hull, were not used, and indeed would have been most inappropriate in this communal venture. In the process of making the canoes new sets of social relations were negotiated and made visible.

On another and more abstract plane of meaning the canoes are a dynamic symbolic manifestation of the strength and *power* of the past in the present. But this interest in the past by Wala islanders is not for its own sake. It is stimulated by a desire to direct this power of the past metaphorically, to steer a course for the future. In a canoe one does not travel alone. A canoe is only ultimately important in terms of the wealth and relationships it elicits. The evident pride the men took in the traditional form of the canoes was all about the way these artefacts would empower the community to change their lives. The *kastom* in the canoe is thus a medium for the social and material regeneration of the local community today.

The outward appearance of the canoes constructed today on Wala is similar to those built one hundred years ago. The primary symbolic directionality of canoe movement has always been *towards* Wala rather than away from the island. In the manufacture of these canoes a decision has to be made on which side to attach the outrigger. The prevailing trade winds blow from the south-east, or to the left of Wala as one stands on the beach at Serser, facing the sea. Placing the outrigger on the left-hand side of the canoe means that it is easier to travel to the mainland, because the outrigger breaks the force of the waves. Virtually all canoes, however, have their outriggers attached to the right, making it easier to reach Wala. Emphasis is thus placed on the fleetness and buoyancy of the canoe on the journey home, of their return laden with goods. Canoes have always transported sustenance, wealth and fame to Wala island. Today, they no longer carry tuskers for the *Maki* ceremonies. Instead, they bring sacrificial cargo of a rather different kind: tourists are substituted for tuskers, dollars for ivory.

References

Battaglia, D. (1990), *On the Bones of the Serpent: Person, Memory and Mortality in Sabarl Island Society*, Chicago: University of Chicago Press.

Bonnemaison, J. (1994), *The Tree and the Canoe: History and Ethnogeography of Tanna*, Honolulu: University of Hawaii Press.

Bregulla, H. (1992), *Birds of Vanuatu*, London: Anthony Nelson.

Butler, J. (1990), *Gender Trouble: Feminism and the Subversion of Identity*, London: Routledge.

Clausen, R. (1960), 'Slit drums and ritual in Malekula, New Hebrides', in *Three Regions of Melanesian*, New York: Museum of Primitive Art, University Publishers.

Codrington, R. (1881), 'Religious beliefs in Melanesia', *Journal of the Anthropological Institute*, XXVI: 261–316.

Forge, A. (1979), 'The problem of meaning in art', in S. Mead (ed.), *Exploring the Visual Art of Oceania*, Honolulu: University of Hawaii Press.

Haddon, A. (1937), *Canoes of Oceania*, Vol. 2. Honolulu: Bernice P. Bishop Museum Special Publication 28.

Harrison, P. (1983), *Seabirds: An Identification Guide*, London: Croom Helm.

Jolly, M. (1982), 'Birds and banyans of south Pentecost: Kastom in anti-colonial struggle', *Mankind*, 13: 338–56.

—— (1991a), 'Soaring hawks and grounded persons: the politics of rank and gender in north Vanuatu', in M. Godelier and M. Strthern (eds), *Big Men and Great Men*, Cambridge: Cambridge University Press.

—— (1991b), 'Gifts, commodities and corporeality: food and gender in South Pentecost, Vanuatu', *Canberra Anthropology*, 14: 45–66.

Keesing, R. (1987), 'Anthropology as an interpretative quest', *Current Anthropology*, 28: 161–76.

Keller, J. (1988), 'Woven world: neotraditional symbols of unity in Vanuatu', *Mankind*, 18: 1–13.

Layard, J. (1942), *Stone Men of Malekula*, London: Chatto and Windus.

—— (1972), *The Virgin Archetype*, Zurich: Spring Publications.

—— (n.d.), 'Canoes', unpublished manuscript, John W. Layard Papers, MSS 84. Mandeville Special Collections Library, University of California, San Diego.

MacKenzie, M. (1991), *Androgynous Objects: String Bags and Gender in Central New Guinea*, Melbourne: Harwood Academic Press.

Munn, N. (1977), 'The spatiotemporal transformation of Gawa canoes', *Journale de la Société des Océanistes*, 33: 39–53.

—— (1986), *The Fame of Gawa*, Cambridge: Cambridge University Press.

Somerville, B. (1894), 'Ethnological notes on the New Hebrides', *Journal of the Anthropological Institute*, XXIII: 363–93.

Speiser, F. (1990), *Ethnology of Vanuatu: An Early Twentieth Century Study*, Bathurst: Crawford House Press.

Strathern, M. (1988), *The Gender of the Gift*, Berkeley: University of Chicago Press.

—— (1990), 'Artefacts of history events and the interpretation of images', in J. Siikala (ed.), *Culture and History in the Pacific*, Helsinki: Finnish Anthropological Society Transactions No. 27.

Thomas, N. (1991), *Entangled Objects: Exchange, Material Culture and Colonialism in the Pacific*, London: Harvard University Press.

—— (1995), *Oceanic Art*, London: Thames and Hudson.

Tilley, C. (1991), *Material Culture and Text: The Art of Ambiguity*, London: Routledge.

Weiner, A. (1992), *Inalienable Possessions*, Berkeley: University of California Press.

Wheatley, J. (1992), *A Guide to the Common Trees of Vanuatu*, Port Vila: Department of Forestry.

–3–

The Anthropology of Art
Susanne Küchler

Introduction

Of all aspects of social life, anthropologists are perhaps most reticent about art. The globalization of mass media, of travel and entertainment have created a diverse terrain with blurred borders and de-territorialized identities. Islands of history and of culture associated with the Renaissance origins of western disciplinary concerns with art appear no longer relevant and capable of serving as categories of study and theory building. Fred Myers's (1994) call to cease the privilege of 'thinking in essences' and Alfred Gell's (1992) insistence on 'methodological philistinism' resonate a deeply felt unease across disciplines concerned with the study of art about the epistemological purchase of the category (Stafford 1999, Mitchell 1996, Docherty1996).

This, however, was not always the case. While early twentieth century anthropological writings on art were concerned with aiding the classification of ethnographic collections, within the American tradition it was Nancy Munn's (1966, 1972) pathbreaking work on *Walbiri Iconography* that formulated the possibility of re-analysing seemingly abstract visual systems in relation to culturally specific and visual ways of knowing. In Britain, anthropological work on art was inspired by Anthony Forge's (1970, 1974) writing on Sepik River mens-house painting which forecast much later work in cognitive anthropology (Whitehouse 1992, Losche 1995).

The theoretical context in which these early studies were conceived can be easily described: with notions which developed out of post-structuralist readings of Freud and Benjamin, Bergson and Proust, which claim an underworld of consciousness and subject-centred philosophy: a supra-human memory whose mechanism is used but not controlled by subject-centred remembering, internalised perhaps, but in a forgetting, self-forgetting sense which was seen to surface in artworks. Recent attempts to reclaim the need for an anthropology of art, such as Alfred Gell's (1998) study of *Art and Agency*, resonated concerns that arose out of a post-linguistic, post-semiotic shift towards a picture's function and intentionalities. It was chiefly W.J.T. Mitchell's (1986) study of *Iconology: Image, Text and Ideology* which challenged our belief that language and reading are of higher quality than viewing, spectating and experiencing pictures.

As anthropology partook of the wider philosophical, art historical and art critical discourse, allowing some of its finest studies to achieve 'cutting edge' status, it also fell victim to definitions of art that were exterior to anthropology. It was only with *Art and Agency* that the possibility of a theoretical definition of art within anthropology could be envisioned, one that places objects in a particular relation to persons. Rather than defining the art object in a way satisfactory to aestheticians, philosophers or art historians, 'anything whatsoever could, conceivably,' said Gell, ' be an art object from the anthropological point of view' (1998: 7). His writing (c.f. 1996) captures a theoretical framework for the recognition that any certainty (cf. Danto 1988) about what counts as an artwork is misconceived.

Dismantling such certainty constituted much of the writings on the anthropology of art during the 1990s, from Christopher Steiner's (1994) work on *African Art in Transit* which showed that there can be no claim on territorially bound artistic production to Fred Myers's (1995) work on Australian Aboriginal art and anthropology's role in the mediation of art relations. While much of the earlier anthropological writing about art stressed the overlap between geography, culture and identity, increasingly a more mobile geography is stressed, the fragility/fluidity of cultural boundaries is noted, and positions of artists in global structures of criticism and consumption are recognized (Rogoff 2000). Related to the emergence of global spaces of art production and consumption is an increasing attenuation of authorship: should the questioning of the singularity of identities result in those identities becoming more freely available to others? Can there be an indigenous art history (MacMaster 1999) and who has the right to speak out on what such an art history is or should be (Enwezor 1997) are some of the more pressing questions which are asked.

In an emerging post-colonial world, the self-critical historical investigation of the history of ethnographic collections has dominated anthropological work on art over the last two decades (Clifford 1988, Price 1989;, Errington 1998, Schildkraut and Keim 1998, Karp and Levine 1991). Yet it was Nick Thomas's (1991) work on *Entangled Objects* which signalled a turn away from institutional concerns over the status of ethnographic collections to a concern with an artefact's 'promiscuous' intentions. Instead of assuming that anthropological research of artworks can open up meanings that are as it were 'hidden' in the history of a collection or the formal properties of artworks, artworks were shown to attract multivariate responses as they are moving across cultural and transactual domains, while impacting on social relations in their wake.

It is now a truism to say that social anthropologists have continued to reduce material artefacts to the relations or meanings in which they are embedded; their interpretations treat the object as no more than an illustration of the things that are external to it. Yet no other debate has thrown up as much controversy over what to do with the object in anthropological analysis as that of aesthetics. The aesthetic turn in the anthropology of art was itself motivated by a post-structuralist reading of cultural form. While insisting on the multivocality of authorship, earlier studies coming out of West African ethnography were fatally coupled with a residual insistence on narrative and linear modes of analysis (Jules-Rosette 1984, cf. Carroll 1987). Howard Morphy's (1992, 1991, 1989) call for a cross-cultural study of aesthetics appeared to shed the linguistic paradigm as his analysis focused on the emotive and physiological associations of cross-hatching in Australian Aboriginal bark-painting. It was soon followed by an anthropological study of the 'cultural eye' based on Jeremy Coote's (1992) study of the poetics of colour terms attached to Nilohic cattle. Much of this work was influenced by a move away from romantic, transcendental ideals about the Kantian disinterested aesthetic to a self-involved, often corporeal response to works of art. Susan Buck-Morss (1997) famously establishes the body as the primary locus of cognition through which the world and consequently art is apprehended. She makes an attempt to erase the anaestheticization of modern art as outlined by Walter Benjamin. Aesthetics appeared thus not part of an exterior, outer reality, but as an entity integral to every aspect of social life.

Alfred Gell's essay (1992) on 'The Technology of Enchantment' foreshadows some of the larger arguments that came to the fore in the introduction to his (1998) *Art and Agency*. In that paper, Gell provocatively claimed that the anthropology of art had got virtually nowhere thus far, because it had failed to dissociate itself from projects of aesthetic appreciation, that are to art as theology is to religion. He argued that if the discipline was instead to adopt the position analogous to that of the sociology of religion, it needed a methodological philistinism, equivalent to sociology's method-ological atheism. This anti-aesthetic stance requires one to resist what one can describe as the 'art cult' to which anthropologists engaged in the terms and technicalities of the art world generally subscribe (Marcus and Myers 1995, Gell 1995, Myers 1994). By desisting an aesthetic perspective on art Gell did not, however, advocate a demystifying sociological analysis that would identify the role of art in sustaining class cultures, or in legitimizing dominant ideologies, as this approach was seen to fail to engage with the art objects themselves, with their specificity and efficacy.

Gell's posthumously published *Art and Agency* constitutes a watershed in the anthropology of art as it breaks not just with the legacy of semiotic analysis, but with the lasting assumption that the anthropology of art concerns itself with the 'non-western' system of art appreciation. Gell was relatively uninterested in the questions raised by art world institutions, believing instead that the anthropology of art should address the workings of art in general. This theory should not just be concerned with the definition of the art object, indeed of 'art' itself, but present an anthropology theory instead of continuing to borrow from sociology, history of art and art criticism. Gell advocates a Maussian basis to an anthropological theory of art that focuses on the ostensibly 'peculiar relations between persons and "things", which somehow "appear as", or duty as, persons.'(1998: 9). A biographical and relational perspective supplies the operative frame from which to introspect the moment at which mere things appear person-like and thus constitute a relational context. Implicit is a categorical rejection of the linguistic analogies that have been crucial to many semiotic and symbolic theories of art. No longer is it axiomatic that art is about meaning and communication. Instead, *Art and Agency* suggests that it is about doing.

Shedding the legacy of logocentrism, the question Gell puts to us is not 'how can art be the object of discursive thought? but how can art be 'thought-like', or 'how can thought conduct itself in art'. The anthropology of art outlined by Gell is constructed as a theory of agency, or of the mediation of agency by indexes, It is the formal complexity, or indeed technical virtuosity displayed by indexes, exemplified by apotropaic pattern, which is given central position in this theory. It is crucial to this theory that indexes display a 'certain cognitive indecipherability,' that they tantalize and frustrate the viewer who is unable to recognize at once 'parts and wholes, continuity and discontinuity, synchrony and succession.' Abstract visual systems are shown to be not just 'realist', but to engage certain visual ways of knowing and thinking (Stafford 1996). The recognition that artworks are vehicles of thought, best exemplified by Rosalind Krauss's (1985) brilliant exposition of concepts of originality and newness in Euro-American art, allows Gell to realize the nexus of modernism and the anthropology of art, thus overcoming the latent divide between 'western' and 'non-western' art.

Like Rodin's casts, shown by Krauss to serve as material analogue to the self-organizing capacity of the avant-garde artist, most artworks that have for long been the concern of anthropology are reproductive as well as generative in nature. Artworks such as Australian Aboriginal bark and acrylic paintings trace the workings of intellect-ual economies, where proprietary rights are extended not to objects, but to knowledge based resources (Harrison 1995). It is access to these resources which is hotly con-tested and to which claims are made through the knowledge and reproduction of images in artworks (Morphy 1995).

Artworks that exist as intellectual property are by their very nature readily transmittable, even portable in a more or less tangible sense. As the 'landscapes' of group identity are no longer familiar anthropological objects, as groups migrate, regroup in new locations, reconstruct their histories and reconfigure their ethnic projects, and as global tourism and design takes hold of ever more localized material worlds, the relation between people and things has become 'slippery.' What we may call 'the piracy of identity' (Harrison 1999) through the contested appropriation of cultural imagery has in fact become one of the most public occasions in which art-works achieve notoriety (Brown 1998). It is 'art without borders' which raises the question of whether culture can be copyrighted, for whom and by whom, as well as how the descriptive practice of anthropology should respond. Far from having to be laid to rest, it is now more than ever that an anthropological theory of art is needed.

References

Brown, M. (1998), 'Can Culture be Copyrighted?' *Current Anthropology* 39 (2): 193–222.

Buck-Morss, S. (1997), 'Aesthetics and Anaesthetics: Walter Benjamin's Artwork Essay Reconsidered', in R. Kraus, A. Michelson, Y.-A. Bois, B.H.D. Buchloh and H. Foster (eds), *October: The Second Decade 1986–1996*, Cambridge, MA: MIT Press, 375–413.

Carroll, D. (1987), *Paraesthetics: Foucault, Lyotard, Derrida*, Berlin: Methuen.

Clifford, J. (1988), *Predicament of Culture: Twentieth Century Literature, Ethnography and Art*, Cambridge, MA: Harvard University Press.

Coote, J. (1992), 'Marvels of Everyday Vision: The Anthropology of Aesthetics and the Cattle Keeping Nilotes', in J. Coote and A. Shelton (eds), *Anthropology and Aesthetics*, Oxford: Oxford University Press, pp. 245–75.

Danto (1988), 'Artefact and Art', in *Art and fact: African Art in Anthropological Collections*, New York: Centre for African Art and Presteel, pp. 18–32.

Enwezor, O. (1997), 'Reframing the Black Subject: Ideology and Fantasy in Contemporary South African Representation', *Third Text* 40: 21–40.

Errington, S. (1998), *The Death of Authentic Art and Other Yales of Progress*, Berkeley: University of California Press.

Forge, A. (1970), 'Learning to See in New Guinea', in P. Mayer (ed.), *Socialisation: The Approach from Social Anthropology*, London: Tavistock, 23–31.

—— (1974), 'Style and Meaning in Sepik Art', in A. Forge, *Primitive Art and Society*, London; Routledge, 169–92.

Gell, A. (1992), 'The Technology of Enchantment', in J. Coote and A. Shelton (eds), *Anthropology, Art and Aesthetics*, Oxford: Oxford University Press, 40–67.

—— (1995), 'On Coote's Marvels of Everyday Vision', *Social Analysis* 38: 18–30.

—— (1996), 'Vogel's Net: Traps as Artworks and Artworks as Traps', *Journal of Material Culture*, 1(1): 15–39.

—— (1998), *Art and Agency: Towards a Theory of Art in Anthropology*, Oxford: Oxford University Press.

Harrison, S. (1995), 'Anthropological perspectives on the management of knowledge', *Anthropology Today* 11(5): 10–14.

—— (1999), 'Identity as a Scarce Resource', *Social Anthropology* 7(3): 239–51.

Jules-Rosette, B. (1984), *Messages of Tourist Art: An African Semiotic System in Comparative*, London, New York: Perspective Plenum Press.

Karp, I. and Levine, S. (1991), *Exhibiting Cultures: The Poetics and Politics of Museum Display*, Washington DC: Smithsonian Institution Press.

Krauss, R. (1985), 'The originality of the Avant-Garde and Other Modernist Myth', in R. Krauss, *The Originality of the Avant Garde*, London: Routledge.

Losche, D. (1995), 'The Sepik Gaze: Iconographic interpretation of Abelam Form', *Social Analysis* 38: 47–60.

MacMaster, G. (1999), 'Towards an Aboriginal Art History', in J. Rushing III (ed.), *Native American Art*, London: Routledge, 81–97.

Marcus, G. and Myers, F. (1995), *Traffic in Culture*, Berkeley: University of California Press.

Mitchell, W.J.T. (1986), *Iconology: Image, Text, Ideology*, Chicago: University of Chicago Press.

—— (1996), 'What Pictures Really Want', *October* 77: 71–82.

Morphy, H. (1989), 'From Dull to Brilliant: The aesthetics of spiritual power among the Yolngu', *Man* (N.S.) 14(1): 21–41.

—— (1991), *Ancestral Connections*, Oxford: Oxford University Press.

—— (1992), 'Aesthetics in cross-cultural Perspective', *Jasso* 23(1): 1–17.

—— (1995), 'Aboriginal Art in a Global Context', in D. Miller (ed.), *Worlds Apart*, London: Routledge, 211–40.

Munn, N. (1966), 'Visual Categories: An Approach to the Study of Representational Systems', *American Anthropologist* 68: 936–50.

—— (1972), *Walbiri Iconography*, Harmondsworth: Penguin Books.

Myers, F. (1994), 'Beyond the Intentional Fallacy: Art Criticism and the Ethnography of Aboriginal Acrylic painting', *Visual Anthropology Review*, 10: 10–43.

—— (1995), 'Representing Culture: The Production of Discourses for Aboriginal Acrylic Art', in G. Marcus and F. Myers *Traffic in Culture*, Berkeley: University of California Press, 55–95.

Price, S. (1989), *Primitive Art in Civilized Places*, Chicago: University of Chicago Press.

Rogoff, I. (2000), *Terra Infirma: Geography's Visual Culture*, London: Routledge.

Schildkraut, E. and Keim, C. (1998), *The Scramble for Art in Central Africa*, Cambridge: Cambridge University Press.

Stafford, B.M. (1996), *Good Looking: Essays on the Virtue of Images*, Cambridge, MA: MIT Press

—— (1999), *Visual Analogy: Consciousness as the Art of Connecting*, Cambridge, MA: MIT Press.

Steiner, C. (1994), *African Art in Transit*, Cambridge: Cambridge University Press.

Thomas, N. (1991), *Entangled Objects*, Cambridge: Cambridge University Press.

—— and D. Losche (1999), *Double Vision: Art Histories and Colonial Histories in the Pacific*, Cambridge: Cambridge University Press.

Whitehouse, B. (1992), 'Memorable religions: transmission, Codification and change in Divergent Melanesian Contexts', *Man* (N.S.) 27(4): 777–99.

Binding in the Pacific: Between Loops and Knots*

Susanne Küchler

Pacific artworks are well known for their capacity to simultaneously contain and elicit all prior and future works (cf. Wagner 1987, Strathern 1991). As 'multiples,' artworks of this kind counterpoise the western assumption of the unique object and thus engage the conceptual frame of modernism in ways that enable us to overcome the latent opposition of western and non-western art still rampant in anthropological studies of art (cf. Gell 1998). Like Mondrian's grids, such artworks as New Ireland *malanggan* or Trobriand canoe-prows appear to originate from the figural, visualized as a graphic as well as a technical schema which also serves as the discursive element in art that otherwise resists exegesis (Summers 1989). This paper sets out to investigate the figural in the art of the Pacific, as it is exemplified by the design structure and 'cross-hatching' of Australian Aboriginal bark paintings (cf. Morphy 1991), by the looped binding of netbags (MacKenzie 1991) and by my own account of the *malanggan* image to which I will return in this paper (cf. Küchler 1987, 1988, Küchler and Melion 1991).

By revealing the figural in Pacific artworks to be a material trace of alternative techniques of binding, that is of knotting and looping, I hope to shed a new perspective that became known as the 'epidemiology' of cultural representations (Sperber 1985, Gell 1993). Following previous uses of the epidemiological model, the question driving this paper is what enables cultural representations to catch on in certain social and cultural situations and not in others. At present, we have two solutions to such a question: Sperber (1985) argued for susceptibility towards representations to be grounded in the affinity between cognitive processes governing representations and cognitive processes sustaining culture. Gell (1993), conversely, saw susceptibility to be embedded in the biographical and relational space within which technical virtuosity is selectively appropriated. Binding, in its continuous and yet varied articulation across the Pacific, serves as a perfect example to both re-examine these positions and to point to a possible resolution to the conflicting emphasis on the cognitive versus the social in the validation of cultural representations. This is because binding is not part of an esoteric knowledge technology,

* *Oceania*, Vol. 69(3), 1999, pp. 145–56.

but is profoundly embedded in the mundane and thus habituated space within which knowledge is externalized; in addition to its embeddedness in the mundane, binding is uniquely bringing to bear upon material affective, personal, as well as mathematical and thus cognitive attitudes and concepts which amplify ways of thinking. Rather than assuming that the figurative expresses ideas which are already existing, could it be that its transmission has a dynamic which creates a lasting difference in culture and society?

Rethinking 'Templates'

It has become common practice to describe the figural element of art in the Pacific with the term 'template' (cf. Morphy 1991, Scoditti 1990). 'Templates' are taken to capture the technical and material means which secure the generativity and reproductivity of artworks and to involve geometric and mathematical specific-ations of some kind, such as the graphic layout and geometric design of 'fore- and 'back-ground' in Australian Aboriginal barkpainting, or the mathematically precise refiguration of the body in Kula canoe-prows. The problem with the term template is that it is drawn from mechanical modes of reproduction, notably the printing press and as such embraces notions of linear, homogeneous, separate and local frames of space-time symptomatic of Newtonian physics. The mechanistic world-view, however, is at odds with the 'organic' description of space-time which appears adequate in capturing Melanesian conceptions of events as inseparable, observer contingent and process dependent, and as framed by non-linear, hetero-geneous, multi-dimensional space-times (cf. Munn 1986, Strathern 1990).

Indeed, such an 'organic' view of the universe is not just an adequate description of Pacific concepts of space-time, but given validity in the West with the advent of Einstein's relativity theory. Einstein's world broke up Newton's universe of absolute space and time into a multitude of space-time frames each tied to a part-icular observer, who therefore, not only has a different clock, but a different map. Quantum theory, moreover, demanded that we stop seeing things as separate solid objects with definite locations in space and time. Instead, they are delocalized, indefinite, mutually entangled entities that evolve like organisms (cf. Ho 1998: 44, Barrow 1992, Wassmann 1994).

The inadequacy, and should we say 'datedness', of the term template and thus of existing descriptions of the figural in art in terms of a mechanistic description thus hardly needs further elaboration. It suggests that the formal properties of artworks are fixed by a technical code which is learned as an aspect of esoteric knowledge, and implies in its mechanical description an ego-centric and anthro-pomorphic spatial cognition that is at odds with its conception in culture (cf. Wassmann 1994). Yet what term, and what set of assumptions, could replace this obviously misleading term? This question was raised also by the evolving New Genetics which found itself unable to comprehend the structure of DNA as a dynamic and evolving, 'organic' system as long as it relied on the notion of template as descriptive vocabulary. In our search for a new model, we may consider following New Genetics which recently began to look towards the science of topology, a branch of mathematics which developed as computing allowed for the visualization of the behaviour of non-linear, organic space-time. The conceptual key to topology is the geometric constancy of objects under deformation which topology sees best exemplified by the behaviour of the knot.

The appeal of the knot as a model for self-organizing, non-linear systems reverberated over the last twenty years not just in science, but also in the humanities (Piaget and Inhelder 1967, Laing 1970, Lacan 1977, Serres 1995). It is this ubiquitous presence of the knot in contemporary academic thought which, I would argue, shaped the resonances evoked by recent analyses of the figural in Pacific art. I am thinking here of MacKenzie's (1991) acclaimed research on looping technologies in Papua New Guinea and Valeri's (1985) account of binding in Hawaii. These studies point to the fact that the knot is not just the singularly most important means of fastening across the Pacific, but works as knowledge technology which enables the externalization of concepts, from sorcery to affinity, into a spatial medium (cf. Levinson 1992).

When we rethink what still is present as mechanistic accounts of technique in these analyses in terms of a notion of a cultural topology, MacKenzie's (1991: 6) insight into the consistent application of alternative modes of binding – surfacing on the one hand in the loop in Non-Austronesian speaking cultures in the Pacific, and on the other hand in the knot in Austronesian speaking cultures – takes on a new significance. Is it possible, one begins to ask, that the non-random distribution of distinctively bound representations across the Pacific is indicative of diverse models of spatio-temporal relations? Could the spread of binding in its different logical articulations have facilitated the regional integration of diverse politico-religious institutions?

The full answer to these questions would demand a much more extensive excursion into Pacific art and society than is possible in the space of an essay. For this reason I will limit myself to the re-examination of *malanggan* ritual art from New Ireland in Papua New Guinea. By sketching its position in the wider field of looped and knotted representations in the Pacific, questions will emerge that will enable us to rethink the relation between art and society in the Pacific.

Binding in the Pacific

From Polynesia to the Non-Austronesian speaking cultures of mainland Papua New Guinea, binding appears not just as technique resonating in everyday tasks from building to repairing bridges, houses and containers of all kinds, but also as artwork. Yet despite this almost trivial and commonsensical presence of binding, the manipulation of string and the visual and formal properties of products of binding varies from place to place, with the most pronounced differentiation emerging between what we call 'looping' and 'knotting' technologies. Looping involves the pulling of the string through the knot, creating an expandable mesh which draws attention visually and conceptually to the threaded string and its continuous run. Knotting, on the other hand, creates a planar surface which covers the knot. The knot, compared with the loop, is only visible when unravelled, thus

Figure 3.1 Sketch of *malanggan*, British Museum XXXVI

drawing visually and conceptually attention to a negative, absent space. While we might be readily inclined to compare examples of looped string with those of knotted string, carved *malanggan*-figures from the North of New Ireland never before were associated with binding (Fig. 3.1). In fact, however, the liminal position of open-work carving between the loop and the knot enables us to draw attention to the dynamic relation between the figural and the social in Pacific art.

Knots and artefacts composed of knots are of paramount importance to the fashioning of contractual relations across Polynesian island cultures. Famous are Tongan barkcloth (*gatu vagatoga*) figuring prominently in ceremonial gift exchanges as the inalienable property of chiefly lineages; the surface pattern of these mats is the result of the rubbing of bark-cloth over a net spanned across the *kupeti*-board (Kooijman 1977). Yet it is possibly Babadzan's (1993) exposition on the Tahitian *to'o* which illustrates most succinctly the conceptual purchase of artefacts of knotting.

To'o are bindings of sennit cordage and tapa barkcloth, sometimes with roughly delineated facial features and limbs woven onto the outside (Fig. 3.2). At all times, bar the ritual at which *to'o* were revealed to view, such tightly knotted bindings encompassed a pillar that kept apart and thus connected heaven and earth and thus invested *to'o* with the god's presence by virtue of contiguity rather than resemblance. The *to'o*, amalgams of ranked images, were thus embodiments of Oro, the Tahitian god, and control over them was essential to the system of social rank and political power. The correlation between the ranked polity of images and social rank among human beings was given regular and formal expression in a ritual called *pa'iatua* which, translated, means the 'wrapping of the gods' (Babadzan 1993, Gell 1998: 111).

No one has as yet better revealed the importance ascribed in Polynesian culture to binding and to the knotted cord as representation of binding (evidenced in Babadzan's work on Tahiti) than Valeri in his work on the king's sacrifice in Hawaii (1985: 296–300). In Hawaii, a sacred cord (*'aha*) acted as the reference point of genealogy – representing not just the king's relationship with the gods, but also the connecting force of genealogy that 'binds together all other genealogies, since it is their reference point and the locus of their legitimacy and truth.' (Valeri 1985: 296). The cord of Hawaiian kingship was not inherited – the undoing of the King's sacred cord dissolving the social bond embodied by the king. The strands obtained from the undoing of the cords were woven into caskets in which the bones of the king are enshrined (Valeri 1985: 298). During the king's reign, the knotting of the cord which celebrated his installation was re-enacted repeatedly as the central organizing rite of the sacrifice of the king. The metaphoric or real 'twisting' of the strands that make up the *'aha* cord enclosed and thereby removed from sight the space where the knot resides, containing and thus arresting the divine powers which come to form the mystical body of kingship.

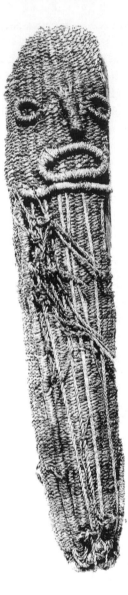

Figure 3.2 *To'o*. God image made from wood with sennit, found in the society islands, Tahiti 1881. British Museum Q81.oc.1550

Across Polynesia, artefacts of binding are thus also a means of wrapping and as such are powerfully associated with contexts and powers of revelation (cf. Gell 1993). Moving to Melanesia, the knot appears to vanish as an artefact of wrapping. This apparent displacement of binding, together with a decrease in emphasis on cloth wealth, appears marked even as one moves from the southernmost islands of the Bismarck Archipelago, such as Vanuatu and the Solomons, to its northern-most extensions in New Ireland and New Hanover. And yet, a much less obvious rendering of the knot as artefact can be found even in those parts of Melanesia that hitherto appeared to lie outside the parameters of Polynesian social systems with their emphasis on the revelatory powers of the wrap and the ranking of lineages.

Where carving from wood predominates in the ritual arts, such as in the North of New Ireland, the knot as the figural frame for technical knowledge appears far-fetched at first glance. Wood is a rigid substance, so much so that it is difficult to imagine a surface carved from wood as having the quality of stretchable rubber which allows one to image the twisting and tangling of string into a knot. That wood indeed lends itself for the technical exploration of knotting is revealed by the American artist Brent Collins whose work visualizes a rather unusual explicit application of knot theory to wood (Fig. 3.3). Many of his pieces display invariant symmetrical relations and uniform thickness from which one can abstract closed, knotted and linked ribbons curving through space (Francis with Collins 1993: 59). The mathematical surface depicted by his artworks indeed constitutes what he calls a 'knot-spanning surface' or a 'framed link,' with the knot literally being carved out of the wood and constituting the hollow, negative space of the sculpture.

The entangling of carved planes, common to Collins' knot-sculptures, is strikingly reminiscent of the figural qualities of *malanggan* from the North of New Ireland (cf. Fig. 3.1). 'Made as skins' (*retak*) to contain the life-force of the deceased persons prior to their final expulsion into the land of the dead, *malanggan* figures are not just visual renderings of knots but partake of the technical vocab-ulary and skill that is common to a range of knotted artefacts such as nets, both fish-nets and nets used as the base of head-dresses worn by *malanggan*-dancers, and plaited artefacts such as mats and baskets. In fact, the northern New Ireland term for knot, *wu-ap*, is associated phonetically and conceptually with the term for an image-based resource from which *malanggan* figures are derived which is *wu-ne*, meaning literally 'connected knots.'

We know New Ireland largely through such *malanggan* artefacts which began to reach us in their many thousands with the early colonial explorations in the mid nineteenth century. Geographically and linguistically, the island forms in many ways a border between the mainland of New Guinea and the island Pacific region. Nearly 400 km long, yet in part only 5 km wide, the island forms the northernmost extension of the Bismarck Archipelago whose volcanic islands dissect in a north-west to south-easterly direction the tectonic plate which connects the island of Papua New Guinea with Southeast Asia. Too far from the mainland to be reached by sea, yet not quite far enough to be seen as an independent entity historically and politically speaking, New Ireland marks the most western extension of the

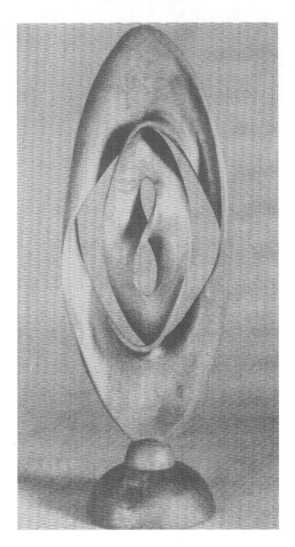

Figure 3.3 Brent Collins, *One-Sided Surface with Opposed Cheiralities*, oiled cedar, 30´12´4 in, 1984

archaeological evidence of the Lapita Pottery culture whose contemporary articulation is found in a concern with tattooing and barkcloth technology that spans most of Polynesia. Though barkcloth is no longer produced on the island today, artefacts made out of barkcloth were collected in the latter part of the nineteenth century reminiscent of today's round, woven *malanggan* known as *warwara*; this fact not only verifies that New Ireland is part of the cloth dominated cultural area of the wider Pacific, but also testifies to an unsuspected relation between *malanggan* and barkcloth technology. Before returning to investigating this relation further, a brief ethnographic sketch of the *malanggan* may help to situate its culture at the cross-roads of East and West.

The Knot-spanning Surfaces of *Malanggan*

To this day, *malanggan* is a rare ethnographic example of a contemporary funerary ritual that culminates in the production, revelation and 'death' of effigies which are part of complex monumental architecture. The effigy when installed in a structure that is specifically erected for its display is thought to be alive, having been gradually animated during the process of its production. Like the effigies of kings in fourteenth and fifteenth century England, which matched or even eclipsed the dead body itself, the effigy is attended as though the dummy was the living person himself – being animated and subsequently allowed to die, thus allowing the deceased person's soul to achieve symbolic immortality.[1] The height of *malanggan* ritual is the dramatic revelation of the effigy inside its architectural structure, followed moments later by the symbolic activation of its death. What took often more than three months to prepare is over in an hour, the 'empty' 'remains' of effigies being taken to the forest to be left to decompose the same afternoon and the architectural structures being removed from sight.[2]

While strikingly similar in the treatment of the effigy as double, this is where any comparison between *malanggan* and renaissance funerary statues must rest. Effigies which are created for the 'finishing of the dead' in the north of New Ireland are neither portraits of deceased persons, nor are they emblematic in character. Instead, these effigies serve as vessels for the transmission of names which are considered to be the source of an augmentable, regenerative life-force. Becoming an ancestor in the *malanggan*-culture, like elsewhere in Melanesia, means to become part of a complex system of names; that is, to become synonymous with a vast resource of knowledge that is almost permanently silenced as the names of those who died are barred from utterance. The peculiarity in the case of *malanggan* is that names are 'found,' validated and transferred as images that are produced out of named 'sources' (*wune*).

Names become skins as a result of being shaped into effigies, but they also become images through the effigy's anticipated death and disappearance from sight. The imagistic character of names in New Ireland enables names to be recalled or 'found' in a controlled fashion using mnemonic techniques which leave their imprint in the form given to effigies. The control authorized over the recall and recognition of ancestral names stands in stark contrast to the image of involuntary remembering which in New Ireland is encapsulated in the image of bush spirits; '*rulrul*' who snatch human skin and appear in their disguise, haunt and trick the living with their sudden and momentous appearances and their seemingly affirmed identity. A *malanggan* effigy, on the other hand, instigates a search which is completed only as a fleeting moment of recognition; to find the name in a sculpted image is possible only by recognizing the effigy as memory trace. The connection between the realm of the ancestral and the realm of the living is visibilized in the figural properties of the effigy and is alluded to in a revelatory manner which we associate with cloth based cultures.

Visually and conceptually, the figures recall a body wrapped in images which draw attention to bodily folds, contours and shape. Incised to the point of breakage, the emerging fretwork takes the forms of instantly recognizable motifs found in abundance in the physical and animate environment of the island culture which produces them. We can recognize in the carved and painted planes birds, pigs, fish and seashells which are depicted with such an accuracy and attention to detail that they appear almost lifelike; the same can be said for the figure set within the fretwork which appears to stare at the beholder with eyes that could hardly be more vivid. Surrealist artists, from well-known figures such as Giacometti to others such as Brignioni, were attracted to these figures however not because of their lifelike character, but because of the apparently ornamental and yet not quite self-evident nature of the shape given to the figures. Motifs appear enchained as figures are standing inside the mouth of rock-cods, framed by many different kinds of fish that bite into limbs and chins, birds that bite into snakes and snakes into birds, and the skulls of pigs that appear to metamorphose into birds. Inner shapes appear enclosed by outer frames in ways which contest the apparent reality of what is depicted like a vision in a dream.

One may be inclined to suggest that these figures look like story boards visualizing myths of potential importance to the understanding of this culture. In fact, such stories exist and are quite readily provided to those connoisseurs or tourists seeking to purchase them as a memento of their island experience. In a search for meaning we direct our eyes away from the hollowed spaces clustering between figure and frame. Yet, one may wonder about these hollowed spaces and indeed, it is here in what is rendered absent through incising that we find a surprising clue of what may count as a description of what a *malanggan* is: what we are looking at are complex knot-spanning surfaces reminiscent of the string-figures which form a beloved pastime across the Pacific.

The finding of a knot in the hollowed spaces of the wood distinguishes ritually effective artefacts from those which are considered 'mistakes' – the space framed by the enveloping planes of the carving calls to mind the heap of wood-chips left at the back of the carvers hut, called *rotap*, literally 'salty rubbish' likened to the dead, uprooted trees which drift ashore once a year from the direction of ancestral lands. The knot which is visible as negative space reflects thus upon a connection with an ancestral domain that is created through ritualized forgetting, one that is kept at bay at all costs (cf. Babadzan 1993).

If a carved knot can in this way be seen to call up what is absent through the creation of negative space, a contrast emerges with the emphasis on life in line, pattern and colour common to string and looping cultures which predominate on the mainland of Papua New Guinea (MacKenzie 1991, Hauser-Schäublin 1996). Hauser-Schäublin described the experience of ancestral power among the Abelam as captured by the brilliance of the line and the initiant's nettle beaten body – an embodiment of ancestral power which appears subversive from the perspective of New Ireland where any claim to personal control over ancestral power is considered deeply ambiguous and possible only under the guise of ritualized acts of negation (cf. Wagner 1987). The knot which, like all wrappings, exists only to be opened calls to mind the impossibility of containing and thus experiencing ancestral power which is effective only as long as it remains distanced and yet 'recollectable' by covering it in secondary 'skins' whose removal, rotting and re-fabrication inscribe the ancestral domain in cycles of movement and arrest.

Knots are found in many contexts in New Ireland culture, all having in common that they serve as a means of 'binding' in more than just a functional sense. There are knotted stems of the gingerplant which are tied around trees in the forest to demarcate land taken out of cultivation for certain periods of time; usually coinciding with the death of a landholding person, the tying and untying of the knot marks the transition of landownership that is associated with the right to taboo (*dang*) the land in this manner. There are also nets knotted across the front of certain houses which are the dwellings of the living dead known as '*haio*' or 'the caught ones,' who incorporate the knowledge that connects the living to the dead. There are nets worn on the head of *malanggan* dancers, alone able to finish the death of a *haio* by recapturing her life-force, who draw with their movements knotted patterns into the sand which disappear as quickly as they appear. And there are knots tied rhythmically into songs classified as *malanggan*, and fishnets which are knotted and used only with the strictest observance of taboos to avoid the splitting of the netting. Not to forget are basketry and mat-work whose plaits interlope in ways recalled by the intersecting planes of sculpted and woven *malanggan* – all made not to invite introspection, but to deflect it.

The negative space of the *malanggan*-figure or the invisible underside or innerlayer of mats and baskets recalled an ancestral realm that is positioned beyond the horizon in a land called *Karoro*. Expelled into the land of the dead, the ancestral remains attached to the living by means of the tangled planes of wood and the twisted leaves of the coconut palm whose seasonal drifting ashore announces the new year of the agricultural calendar. In the same way as magic is needed for the pulling ashore of what is colloquially termed the 'smell' of the dead (*wangam a musung*) in the form of uprooted trees and branches, so the creation of 'skins' carved into wood or bound into coconut leaves which visually and conceptually bind the ancestral to the living cannot be secured by skilled acts alone. The knot which traps the ancestral as it returns to the land of the living has to be found (seen – *kalymi*) in dreams. Being able to artificially induce or otherwise voluntarily effect the dreaming of an image that is able to capture the 'smell' (*ngusung*) of the dead distinguishes carvers (the 'joiners of skin' – *retak*) from others who may own knots in the form of named images, but cannot render them effective.[3]

The contractual relation between the living and the dead that is symbolized by the knotted planes of the *malanggan*-figure, also extends to land. Thus a contrast emerges with the Polynesian material on knotted artworks: while the sennit binding of the Tahitian *to'o* does not appear to have been made to symbolically underwrite a contract over land made between the living, this is certainly the case with *malanggan*. Like the looped netbags of mainland New Guinea, a *malanggan*-figure is always made for someone who stands in a particular social relation to the current owner of the image.

In being made to be sold (*sorolis*), or lent (*aradem*) to affines or even clan-members who may live in different villages or even dialect groups as affirmation of a contract over usufructury rights over land, a *malanggan*-figure moves at least notionally every time it is made. Again like a netbag, a figure connects those who made and those who receive the art-work in that the complex, intertwining planes of a *malanggan*-figure are revealing the memory of contractual relations over land. As rights to land are shared and fluid across linguistic and spatial boundaries, so the right to carve images that evoke such rights are shared; people talk about the 'path' (*selen*) of a named image which flows like a river from a source (*wune*). Such a 'path' is at least partially discernible and reconstructable as *malanggan*-figures document the conceptual and actual fragmentation of images in reproduction as they are shared out into smaller and smaller bits along ever-expanding pathways, only occasionally to be reassembled into larger, more complete assemblages.

There are two techniques of creating 'fractal,' or breakable figures, each associated with distinct contractual relationships: 'breakable' not just in the metaphorical sense as points of breakage suggested by the intersections of knotted planes may be targeted when severing a figure after it served to inform a contract between groups who each take home one part as reminder of the relation thus effected. Each technique makes use of the knot in different ways; one to render different kinds of knots in the planes carved from wood, of which there are three main named types, and the other to trace the deformation of a knot into its possible forms. Elsewhere (Küchler and Melion 1991) I have pointed to the importance of motifs and motif combinations as tracing the spatio-temporal structure of the ritual process itself and marking it as sacrificial in nature: images of absorption foreshadow images of containment and beyond that of decomposition and transcendence. It is, however, the technique of carving surface spanning knots from wood which explains how images are generated and reproduced from an ever-changing and yet fixed repertoire which allows the instantaneous recognition of the named 'source' from which an image is derived. The processuality of ritual itself may be derivative of the process of knotting which suggests the opening, severing and dissolution of whatever has been connected.

This method of creating 'telling' variations in the material forms of *malanggan* is used for the six named image 'sources' of *malanggan*, which are in turn paired into three sets of two. As over 15,000 artifacts have been collected in mainly just six villages over the last hundred years, and the number of actually produced artifacts may be double that, one can easily predict that variations in form will be minute, a suggestion that any collection, however small or scantily sampled, supports. Like the sound of the names which are chanted when a figure is seen for the first and often only time, or the steps which are traced on the ground when a dance is learned, the knotted shape of a figure contracts, expands and entangles to create for a fleeting moment a container for a knowledge that is tantamount to a testimony – you know what you have seen.

In behaving like looped artefacts in underscoring reproductive metaphors and relations, yet visually and conceptually alluding to the necessary distancing of the ancestral by means of knotting, the case of *malanggan*-figures falls awkwardly within the neat opposition of the loop and the knot. Yet it is precisely this non-fit which draws attention to the dynamic which keeps looping and knotting as cultural representations within distinct social and historical fields.

The Knot and the Loop: The Differentiation of Binding in the Pacific

Studies of knots as material renderings of binding in figural, cord-like or patterned form, are surprisingly rare in Oceanic ethnography, despite the prominence of an allied technology known as looping whose conceptual force was recently uncovered by MacKenzie (1991) for the non-Austronesian speaking cultures of the mainland of Papua New Guinea. MacKenzie's study of netbags showed more than any previous ethnography the intimate link between a technique of binding, in this case that of 'looping', and the externalization and management of knowledge. Her study shows netbags to work as a vehicle of ideas that identify spatio-temporal frames with the continuity of the string whose continuous run forms the body of the ubiquitous 'bilum.' Hauser-Schäublin (1996) recently extended this examination of string-based cultural imagination to concerns with the line and the frond in Sepik River figural art, while arguing for an emerging revelatory 'aesthetic' that is diametrically opposed to the emphasis on the plait and the plane in island Melanesia and Polynesia. Even if we do not want to agree with her identification of the string as 'non-cloth' cultural 'aesthetics,' her study is of profound importance for a study of binding in the Pacific as it suggests a complex and yet systematic contrast between looping and knotting as allied and yet mutually opposed knowledge technologies.

Mathematically speaking, the difference between the loop and the knot lies in their execution only, one being literally the flip-side of the other. As Adams summarizes succinctly in his introduction to knot-theory, the knot is a **knotted loop of string**, except that we think of the string as having no thickness, its cross-section being a single point.[4] The knot is then a closed curve in space that does not intersect itself anywhere, the most basic of which is the circle or 'un-knot'. While the knot thus lives in the space visualized by a surface, the loop *is* the surface of intersecting planes.

Returning to my initial considerations of the nature of 'template,' the consideration of the contrast between knotting and looping technologies in the Pacific opens a new, surprising perspective: it appears as if indeed the mechanical description of a template as a technique for reproducing discrete space-time frames within fixed locales is quite apt for an account of looped material culture, yet not for the material renderings of knots, and that we are therefore not dealing with an albeit vanishing, difference between 'us' and 'them', but with different, though mutually interdependent ways of framing knowledge. The contrast between the types of binding is brought into sharper relief by considering the differences in the description of each.

Looped netbags as described by MacKenzie (1991) come in a range of numerous different, yet largely 'fixed' and localized 'styles,' with bags being defined in terms of different functions associated with the size', texture and finish of each item. Single cultural areas such as the Telefol, as described by MacKenzie (1991: 46–51), distinguish twenty-seven named types of netbags, and sub-types which are grouped into two categories (large flexible 'bilums' and small tightly looped 'bilums') and a further five subcategories (mouth-band 'bilum', father 'bilum', pocket 'bilum', twin 'bilum', red seed 'bilum'). The persistence of such categories as well as the techniques which are associated with them broadly frame regional and intra-regional boundaries, as both the bags as well as the technique of looping associated with them are transmitted along affinal paths. As the construction process and the resulting form of the bilum is inseparable from the selected looping technique, a variation in the technique has obvious stylistic consequences. Most important, however, is MacKenzie's observation that technique is learned through observation and imitation, a process that is initiated in early childhood.

In stark contrast to the mechanical description of looping as a *technique* of binding, knotted artefacts are 'seen' and described in terms of figural topologies which emphasize a dynamics of ordering inherent in the patterned design of binding. Given the emphasis on a visually mediated 'knowing' or understanding of designs which is acquired through 'dreaming', the execution of knotted forms is not restricted to those who have acclaimed 'rights' to designs. As we saw in the example of *malanggan*, the ownership of a design can be sold, lent or otherwise transmitted, with one carver being hired across the region. As the technical execution of the design is subordinate to the resulting pattern in these knotted artworks, variation of technique is less frequent and indeed even avoided as this could enhance the possibility that designs are confused and theft allegations are brought forth.

The variability of design in knotted artefacts is thus governed by factors other than the conditions influencing the transmission of skill based technique. In fact, as pointed out earlier with reference to the practice of carving 'fractal' images in New Ireland, the variability of design is a direct consequence of the embeddedness of the design in contractual relations over land which are visually and metaphorically tied into the tangled planes of the *malanggan*-figure.

The 'knotted' planes of *malanggan* figures in New Ireland thus command to be perceived as a geometric configuration in order for execution to produce acceptable results. In comparison with looped netbags, *malanggan* figures come in limited shapes, sizes and motifs with an emphasis on their combination that remains constant throughout the north of the island while allowing for infinite variability at the level of 'surface' design such as in the painted patterns and painted or carved, attached motifs. The stability of pattern is the result of the subordination of technique to the spatial framing of the image. Rather than 'finding' the design of the figure in the wood during carving, the geometric contours of the image are drawn on sand or onto the wood that is used for carving, thus allowing improved or innovative techniques of fabrication only minimal scope to influence the resulting product. It is this geometric design of a *malanggan* which is owned and exchanged in the north of New Ireland, quite unlike the New Guinea netbag described by MacKenzie (1991) which is validated in terms of its technical execution and exchanged as artefact.

The emphasis on the configuration of design rather than its execution also explains the frequently lamented decline in the degree of visual open-work in *malanggan*; noted as negative by the western connoisseur who measures aesthetic appeal in terms of skilful execution, the decreasing three-dimensionality of carving does not bother anyone in New Ireland as even a sketched, more or less painted design would do if time and attention is demanded for tasks other than the preparation of feastings.[5] At the same time, figures are seen as belonging to the same 'source' or knot-type, even if one is executed in wood and the other in fibre and even if place and time of execution are unconnected.

Making such comparisons, which are by definition unverifiable as the artefact is destroyed or sold subsequent to its use in ritual, implies that the experience to which reference is made lies outside the making of ritual artworks in the mundane tasks of binding. Yet should we imagine people walking around with an abstract image of a knot in their heads, its different types and possible deformations, which then just get projected onto a given medium? Or should we imagine that pattern is deduced from the structure inherent in the practical operation of knottings. These quite plausible solutions to the problem of pattern production and recognition ignore an aspect of pattern that has become known as self-organization whose explanation has become crucial to the modelling of non-mechanistic biology (Saunders 1998: 55). Confronted with the evidence of non-linear systems which are capable of self-organization, Lovelock (1989) recently proposed that what appears as self-regulation is a co-ordinated phenomenon arising from causes that are themselves co-ordinated and purposeful. An overall 'design' thus does not have to be in existence, while the system itself has to be inherently co-ordinated and purposeful, ascribing thereby autonomy and agency to non-linear organic systems compared with the dependency on external forces as suggested in the mechanical description of systems.

Lifted out of biology to the world of material culture, the notion of co-ordination and purpose throws a new light on the poignancy of the knot as metaphor of generative systems. The knot might be argued to work as 'source' not because it is comprehended in terms of a quasi pre-existing mathematical and spatially decentred system of ordering, but because it is a trace of purposeful co-ordination. And it is here, in being a trace of **purposeful co-ordination** that the loop and the knot meet, diverging 'merely' in the relative importance ascribed to either technical process or pattern in the generative reproduction of things. Following the insights of new biology, however, each manner of co-ordination, the mechanistic and the organic, would bring with it associated space-time frames which in turn would create a difference in culture and society.

Turning back to the distribution of looped and knotted artworks in the Pacific we note with MacKenzie (1991) the neat overlapping of the two sides of binding with the spread of non-Austronesian and Austronesian speaking cultures. One may be reminded of Whitney Davis's (1986) argument in his 'origin of image-making' that images originate from mark making, not from pre-existing concepts that are 'projected' onto materials. The story of the loop and knot may support his case and provide a new perspective on the epidemiology of cultural representations: one that draws attention to the mundane, affective, personal and yet profoundly mathematical nature of binding which shapes ways of thinking.

Notes

1. E. Kantorowicz (1957), see also C. Ginzburg 1991.
2. Many figures are taken to the nearest mission or purpose built 'custom house' to await a foreign buyer.

3. One is reminded of Howes (1988) poignant insight into the correspondence of a cultural emphasis on olfaction with a conceptual displacement of the dead from the realm of the living and its distinct patterns of distribution across the Pacific. I would add to this the corresponding amplification of knotting as 'cognitive style' common to socio-political institutions that grow out of the 'the finishing' or expulsion of the dead.

4. My understanding of knot theory is almost entirely governed by C. Adams's (1994) extremely enlightening *The Knot Book: An Elementary Introduction to the Mathematical Theory of Knots.*

5. It is, however, noted as a 'sad' consequence of the constraints of modern life as this makes figures much less attractive to the western buyer.

References

Adams, C. (1994), *The Knot Book: An Elementary Introduction to the Mathematical Theory of Knots*, New York: W.H. Freeman and Company.

Babadzan, A. (1993), *Les dépouilles des dieux: Essai sur Ie religion tahitienne à l'époque de la découverte*, Paris: Editions de la Maison des sciences de 10homme.

Barrow, J. (1992), P*i in the Sky: Countin,. Thinking and Being*, Oxford: Oxford University Press.

Davis, W. (1986), 'The origin of image-making', *Current Anthropology* 27: 193–215.

Francis, G. with Collins, B. (1993), 'On Knot-spanning Surfaces: An Illustrated Essay on Topological Art', in M. Emmer (ed.), *The Visual Mind: Art and Mathematics*, Cambridge, MA: MIT Press.

Gell, A. (1993), *Wrapping in Images: Tatooing in Polynesia*, Oxford: Oxford University Press.

—— (1998), *Art and Agency: Towards a New Anthropological Theory*, Oxford: Oxford University Press.

Ginzburg, C (1991), 'Repräsentation: das Wort, die Vorstellung, der Gegenstand', *Freibeuter* 22: 3–23.

Hauser-Schäublin, B. (1996), 'The Thrill of the Line, the String, and the Frond or why the Abelam are a non-cloth culture', *Oceania* 67(20): 81–106.

Ho, M.-W. (1998), 'The New Age of the Organism', *Architectural Design.* No 129: 44–51.

Howes, D. (1988), 'On the Odour of the Soul: Spatial Representation and Olfactory Classification in Eastern Indonesia and Western Melanesia', *Bijdragen Tot de Taal- Land- en Volkenkunde*, 144: 84–113.

Kantorowicz, E. H. (1957), *The Kings Two Bodies: A Study in Mediaeval Political Theology*, Princeton, NJ: Princeton University Press.

Kooijman, S. (1977), *Tapa on Moce Island Fiji*, Leiden: E. Brill.

Küchler, S. (1987), 'Malangan: Art and Memory in a Melanesian Society', *Man* (n.s.) 22(2): 238–55.

—— (1988), 'Malangan: Objects, Sacrifice and the Production of Memory', *American Ethnologist* 15(4): 625–37.

Küchler, S. and Melion, W. (eds) (1991), *Images of Memory: On Remembering and Representation*, Washington, DC: Smithsonian Institution Press.

Lacan, J. (1977), *The Four Fundamental Concepts of Psycho-Analysis*, J.-. Miller (ed.), A. Sheridan (tr.), New York: WW. Norton & Co.

Laing, R.D. (1970), *Knots*, New York: Random House.

Levinson, S. (1992), 'Primer for the Field Investigation of Spatial Description and Conception', *Pragmatics* (1): 5–47.

Lovelock, J.E. (1989), *The Ages of Gaia*, Oxford: Oxford University Press.

MacKenzie, M. (1991), *Androgynous Objects: Stringbags and Gender in the Pacific*, Berlin: Harwood Press.

Morphy. H. (1991), *Ancestral Connections*, Oxford: Oxford University Press.

Munn, N. (1986), *The Fame of Gawa: A Symbolic Study of Value Transformation in a Massim (Papua New Guinea) Society*, Cambridge: Cambridge University Press.

Piaget, J. and Inhelder, B. (1967), *The Childs Conception of Space*, F.J. Langdon and J.J. Lunzer (trs), New York: W.W. Norton & Co, Inc. [1948,1956]

Saunders, P. (1998), 'Nonlinearity: What it is and Why it Matters', *Architectural Design* No 129: 52–7.

Scoditti, G. (1990), *Kitawa: A Linguistic and Aesthetic Analysis of Visual Art in Melanesia*, Berlin: Mouton De Gruyter. At least chs 5–7.

Serres, M. (1995), *Conversations on Science. Culture and Time with Bruno Latour*, R. Lapidus (tr.), Ann Arbor: University of Michigan Press. [1990]

Sperber, D. (1985), 'Anthropology and Psychology: Towards an Epidemiology of Representations', *Man* (N.S.) 20: 73–89.

Strathern, M. (1990), 'Artefacts of History: Events and the Interpretation of Images', *Culture and History in the Pacific,* 25–44. Jukka Siikala (ed.), Suomen Antropologisen Seuran toimituksia, 27. Helsinki: Finnish Anthropological Society.

—— (1991), *Partial Connections*, Savage, Maryland: Rowman and Littlefields.

Summers, D. (1989), 'Form: Nineteen Century Metaphysics and the Problem of Art Historical Description', *Critical Inquiry* 15: 372–407.

Valeri, V. (1985), *Kingship and Sacrifice in Hawaii*, Chicago: University of Chicago Press.

Wagner, R. (1987), 'Figure-Ground Reversal among the Barok', in L. Lincoln (ed.), *Assemblage of Spirits*, Minneapolis Gallery of Art 56–63.

Wassmann, J. (1994), 'The Yupno as Post-Newtonian Scientists: the question of what is "natural" in spatial descriptions',' *Man* (N.S.) 29(3): 645–67.

−4−

Visual Culture
Christopher Pinney

Introduction

The brief discussion of photographic portraiture in India reprinted here is a compressed consideration of the practices which are considered in much greater depth in *Camera Indica* (1997). Both the article and the book are concerned with the disjunction of surface and depth, or exteriority and interiority. If colonial photographic practice attempted to fix identities on Indians which could be read externally, popular Indian postcolonial practice seems bewitched by the difficulty of establishing a relationship between a slippery surface and an unknowable interior. While aspects of this popular practice run counter to the widely accepted (and in other contexts extremely productive) 'ethnosociological' model of Indian persons as fused domains of bio-moral substance, they resonate with a set of emergent concerns within the post-disciplinary space of material culture.

These concerns have engaged a long Platonic and Kantian tradition which has privileged 'interiority' and denied the body. Since when, Barbara Maria Stafford asked, 'does working with surfaces qualify as shallow?' (1996: 7). The answer is (at least in the European tradition) 'since Plato'. If images were seductive shadows, the Idea could only be accessed cerebrally, through language. In the rest of this introductory text I will try and develop a brief history of a new 'visual culture' which attempts to escape the burden of this legacy and to grant images and objects a priority. As befits a post-disciplinary practice, key elements within 'visual culture' emerge not only from art history, but also from a broader repertoire which includes historiographic concerns and work within what might loosely be amalgamated as 'history of culture/cultural theory/history of ideas'.

The historian Carlo Ginzburg, in his attempt in the 1960s to understand the tradition of visual analysis associated with the Warburg Institute, identified what is perhaps the core problem in the study of visual culture in the observation that 'the historian reads into [images] what he has already learned by other means'. For Ginzburg this was a reflection of a 'physiognomic' circularity: just as physiognomists' readings of faces tell us only about the classificatory system that informs the readings (rather than about the relationship between the face and character), so we unwittingly claim to find evidence in the visual that in fact we have discovered elsewhere. To take a famous – and literally physiognomic – example, who can say (by merely looking at their faces) whether Mr and Mrs Andrews, as depicted by Gainsborough, really were callous land-owners (as John Berger suggested) or dispassionate Rousseaueans (as Lawrence Gowing tried to have us believe).

One seeming solution to this is the claim for the indeterminacy of the visual. Think of how differently a set of film stills might look as you queue up to get into the cinema, and how they might look as you leave having seen the film. The filmic narrative (what Eco calls 'syntagmatic concatenation imbued with argumentative capacity') provides a context which can secure a stability of interpretation. The anthropological version of this has been the assumption that objects have 'social lives' and are subject to 'entanglement', positions established within anthropology by Arjun Appadurai (1986) and Nicholas Thomas (1991). The article reprinted here is one example of such an approach and is a fragment from a larger project which stressed what Elizabeth Edwards described as photography's 'polysemous nature, lack of fixity and context-dependent modes of making meaning' (Edwards 2001: 14). Having moved towards a much more phenomenological position in the years since writing this article, I would now incline to a much greater concern with embodiment and a lesser concern with contextual determination. However, the article raises significant problems for a 'cultural' phenomenology.

Often, declarations of objects' and images' emptiness have become a proof for an anthropology committed to the victory of the cultural over the material, and of the discursive over the figural. Such strategies might be seen as an enduring manifestation of the 'linguistic turn', the humanities-wide preoccupation with the arbitrary and conventionalized nature of social meaning (and as I have indicated, this is certainly a plausible frame within which to consider the piece reprinted here). Part of the radicality of the linguistic turn consisted in its critique of neo-Romantic fictions of the auton-omous object and of self-present meaning. However, it might be argued that in its material-cultural incarnation the stress on the cultural inscription of objects and images has erased any engagement with materiality or visuality except on linguistic terms.

It is against this background that W.J.T. Mitchell (1995) identified an emergent 'pictorial turn' as a reaction against the 'linguistic turn', suggesting that it is about time we stopped asking what images can do for us, and time that we started asking 'What do Pictures Really Want?' By treating images (subjunctively) as 'subalterns' we might, Mitchell implies, escape from the seeming inevitability of writing their histories with evidence derived 'by other means'.

This pictorial turn can be given a long history. John Berger's work from the late 1960s onwards can be rightly critiqued for its frequent 'textualism' and desire to 'read' images. However, his contribution through the recognition that most art history was itself in need of analysis and critique has been invaluable. His was also in large part an 'anthropological' analysis of art institutions and the art idea which also dwelt on the peformative work of images and many current developments still feel like minor modifications in the wake of his poetic critical vision.

Also emerging from art history, the work of Norman Bryson and Michael Fried produced vastly more detailed analyses of the visual force-fields that bind people to images and position images as actors. Their concerns with the complexity of visual regimes and aspects of gazes and glances (through bodies that were variously acknowledged or 'decarnalized') helped theorize a key set of issues for those concerned with the phenomonological dimensions of the beholders of images.

A more explicit theoretical rejection of 'decarnalized' approaches has been elaborated by Barbara Stafford in her bravura critique of what she calls 'cultural textology'. In her visualist manifesto *Good Looking* she fiercely attacked semiotic, deconstructionist and interpretivist approaches for 'reconc[eiving] the material subjects of their inquiry as decorporealized signs and encrypted messages requiring decipherment . . . [reducing] communication to inscription' (1996: 6).

The central problem, Stafford argues, lies in the 'totemization of language as a godlike agency in western culture': we might be able to see, and feel the power of images but still we desire linguistic elaborations of what they really 'mean'. In part this reflects the deep Platonic disparagement of the surface in favour of an interiorized conceptualism, but twentieth-century post-Saussurean approaches laboured under the burden of a linguistic arbitriness whose fluidity was incompatible with the weight of the flesh of the world. 'Saussure's schema', Stafford records, 'emptied the mind of its body' and even 'the so called interdisciplinary "visual culture" programs are governed by the ruling metaphor of reading. Consequently iconicity is treated as an inferior part of a more general semantics' (1996: 5).

For anthropologists, part of this critique will resonate with Dan Sperber's description of 'cryptological' approaches to symbolism, and indeed with the wider debate about the performativity (rather than referentiality) of ritual. What is needed in other words are not theories of representation so much as engagements with embodiment.

The relationship of the human sensorium to different political orders and regimes of taste implicit in Bryson and Fried's work has been most fully developed in Susan Buck-Morss's working through of Walter Benjamin's 1930s insights into the question of embodiment. If one was to construct a long history of visual culture, Benjamin (together with Martin Heidegger) would perhaps emerge as the key figure. Buck-Morss has laid an analytical pathway towards the recuperation of the corporeality of aesthetics. Arguing for the necessity of a recognition of aesthetics' earlier sense as *aisthitikos* (denoting 'perception by feeling'), Buck-Morss mounts a sustained critique of the Kantian abolition of the phenomenal through its privileging of 'disinterest' (a 'decarnalization' to use Bryson's term). This position which resonates with Bourdieu's argument in *Distinction*, is elaborated in a more ethnographic context by David Morgan who, in a study of US Christians reveals how devotees invoke a non-Kantian phenomenology to ease their suffering. His study, replete as it is with accounts of how certain images of Christ are chosen because (say) they are profiles which will allow the devotee to whisper into his ear, or (say) feature a direct visual projection which will exude a protective veil when hung over a child's bed, adumbrates a complex popular embodied 'kitsch' in which what matters in the adjudication of an image's worth is not a disinterested aesthetic, but practical efficacy.

It is increasingly clear that the rise of a disinterested art at the expense of a stigmatized kitsch, and the rise of visuality as 'cultural textology' have been inseparable historical processes. The corollary of this is that the emergence of visual culture as (in part) a repudiation of the neo-Romantic art idea (i.e. art as transcendance) must also involve the re-embracing of the body. In addition to the authors mentioned above, other key figures who are pioneering a new embodied analysis of visual culture include the ethnographic film-maker David MacDougall whose exploration of the filmic implications of Merleau-Ponty's insights have a profound relevance for all anthropologists, and the art historian Alexander Nemerov whose recent interpretation of the 'sensuous identifications' in the work of the nineteenth-century American painter Raphaelle Peale may well come to have a general impact on the humanities' engagement with the visual in a way that parallels that of Berger's *Ways of Seeing* in the early 1970s.

As the twenty-first century unfolds it also becomes increasingly plausible that the entire problematic of an abstracted 'visual' might be usefully understood as an outcome of a particular mode of picturing the world. This is the position argued by Martin Heidegger in his seminal 1935 essay 'The Age of the World Picture', which ambitiously sketches a history of man's alienation from the 'flesh' of the world. By being able to take a 'view' of the world, man freed 'himself to himself' as a locus of agency distinct from the zone of the 'non-human'. Heidegger invoked the term 'picture' (*Weltbild*) to describe modernity's construction of the world to mark a distance between man and the world. 'Picture' stands for something displaceable, substitutable, disposable, the withered sign of a once vital phenomenological presence which impressed itself on man. Martin Jay uses the term 'Cartesian perspectivalism' to describe something very similar to the age of the world as picture, and it would be very easy to assimilate the historical analyses of figures as diverse as Camus, Senghor and Foucault to a trajectory marked by an increasing coldness of vision as a product of the elongated distance between subjects and objects.

A Heideggerean stance is evident also in the work of some producers of 'visual culture' with which anthropologists should be familiar. The Australian Aboriginal photographer Leah King-Smith, for instance, has described how she was troubled by the easy distance that could be established between small, almost disposable archival images of the Australian genocide (which she was employed to work with in the Royal Museum of Victoria in Melbourne). This distance produced an alibi, or an insulating sheen, in the face of the viewers' complicity.

In her *Patterns of Connection* series King-Smith morphed these archival images with contemporary landscapes in enormous cibachrome prints whose physical presence and glossy finish collapsed that distance between the image and the beholder. Rather than insulating the viewer, the glossy surface acted as a partial mirror revealing the presence of the beholder, collapsing distance and drawing him or her into the picture itself so that a true leakage of identity could occur.

Leah King-Smith's desire to engage not simply the disembodied eye of the beholder but his or her corporeal presence (mirrored in the surface of her reflective images) suggests the inevitable slippage of the 'visual' into a broader domain. Visual culture is a necessary, and welcome, post-disciplinary practice. But just as it names a move beyond the concerns of art history, anthropology, etc, it already needs to be superceded by an engagement with embodied culture or performative culture (i.e. culture considered through its fundamental materiality) that recognizes the unified nature of the human sensorium. This need not involve a new nomenclature, merely a recognition that the 'visual' does not mark an exclusive preoccupation with the ocular but denotes, rather, a strategic opposition to everything 'not-visual'. Thus 'visual anthropology' is increasingly recognized not as an exclusive concern with visuality, but a rejection of purely textual/semiotic approaches, which might encompass (via the strategic agency of the 'visual') the auditory, the haptic, etc.

Inasmuch as the visual in this broader definition stands in opposition to the analytically closed domain of certain kinds of texts, Lyotard's distinction between 'discourse' and 'figure' may yet prove useful. Lyotard uses these terms to connote on the one hand 'linguistic-philosophical closure' (discourse) and on the other a domain where 'meaning is not produced and communicated, but intensities are felt' (figure). 'Figure' connotes a densely compressed performative/illocutionary, affectively and libidinally charged domain which escapes conventional signification (see Carroll, 1987).

The manner in which these terms transcend the narrowly 'visual' and 'textual' permits us to see that certain visual forms (e.g. diagrams) are extremely closed, while certain textual forms (e.g. certain kinds of poetry) may be open. The visual per se is not what demands new kinds of analysis, rather it is the visual as part of a broader domain of the 'figural'. It is here that, once again, the 'pictorial turn' reconnects with the broader trajectory of current developments within the study of material culture.

References

Appadurai, A. (ed.) (1986), *The Social Life of Things*, Cambridge: Cambridge University Press.

Benjamin, W. (1992), 'The Work of Art in the Age of Mechanical Reproduction', in *Illuminations*, London: Fontana, 211–35.

Berger, J. (1972), *Ways of Seeing*, Harmondsworth: Penguin Books.

Bryson, N. (1983), *Vision and Painting: the Logic of the Gaze*, New Haven: Yale University Press.

Buck-Morss, S. (1992), 'Aesthetics and Anaesthetics: Walter Benjamin's Artwork Essay Reconsidered', *October* 62: 3–41.

Carroll, D. (1987), *Paraesthetics: Foucault, Lyotard, Derrida*, London: Methuen.

Edwards, E. (2001), *Raw Histories: Photographs, Anthropology and Museums*, Oxford: Berg.

Fried, M. (1980), *Absorption and Theatricality: Painting and Beholder in the Age of Diderot*, Berkeley: University of California Press.

Ginzburg, C. (1989), 'From Aby Warburg to E.H. Gombrich: A Problem of Method', in *Clues, Myths and the Historical Method,* Baltimore: Johns Hopkins University Press.

Heidegger, M. (1977), 'The Age of the World Picture', in *The Question Concerning Technology and Other Essays*, New York: Harper.

Jay, M. (1988), 'Scopic Regimes of Modernity', in Hal Foster (ed.), *Vision and Visuality*, Seattle: Bay Press, 3–27.

Khare, R.S. (1993), 'The Seen and the Unseen: Hindu Distinctions, Experiences and Cultural Reasoning', *Contributions to Indian Sociology* (n.s.) 27(2): 191–212.

King-Smith, L. (1992), *Patterns of Remembrance*, Australian Centre for Photography, Sydney and Southeast Museum of Photography, Daytona Beach, Florida.

MacDougall, D. (1998), *Transcultural Cinema*, Princeton: Princeton University Press.

Marriott, M. (1976), 'Interpreting Indian Society: A Monistic Alternative to Dumont's Dualism', *Journal of Asian Studies* 36, 3.

Mitchell, W.J.T. (1995), 'The Pictorial Turn' in *Picture Theory: Essays on Verbal and Visual Representation*, Chicago: University of Chicago Press, 9–34

—— (1996), 'What Do Pictures Really Want?' *October* 77, 71–82.

Morgan, D. (1998), *Visual Piety: a History and Theory of Popular Religious Images*, Berkeley: University of California Press.

Nemerov, A. (2001), *The Body of Raphaelle Peale: Still Life and Selfhood, 1812–1824*, Berkeley: University of California Press.

Pinney, C. (1997), *Camera Indica: the Social Life of Indian Photographs*, London: Reaktion.

Sperber, D. (1975), *Rethinking Symbolism*, Cambridge: Cambridge University Press.

Stafford, B.M. (1996), *Good Looking: Essays on the Virtues of Images*, Cambridge, MA: MIT Press.

Thomas, N. (1991), *Entangled Objects: Exchange, Material Culture and Colonialism*, Cambridge, MA: Harvard University Press.

Photographic Portraiture in Central India in the 1980s and 1990s*

Christopher Pinney

The whereabouts of the town of Nagda is most easily described in terms of its position mid-way between Bombay and Delhi on the Western Railway broad-gauge line. Nagda has six main photographic studios and it is here that the portrait photographs discussed in this chapter were collected. Precisely parallel to the railway line runs Jawahar Marg which takes one from the station, along the southern flank of the bazaar section of Nagda. About 500 yards from the station, past numerous shops selling tea, clothes, bicycles, hardware and irrigation pumps, is the Venus Studio. Run by two Brahman brothers from the nearby sacred centre of Ujjain, there is a good selection of promotional material on display in their shop front. On most days a life-size cardboard cut-out of a European woman in a tennis skirt holds a box of Konica film and invites customers to enter their premises. Inside, displayed along with samples of their studio work, are a number of very fine prints distributed by Midas Colour laboratories who compete for most of the local colour printing. Some depict glamorous wealthy women of the type normally only to be seen on the screen of the nearby Prakash Talkies. Others show equally beautiful women adorned as village belles, in elaborate quasi-tribal jewellery and embroidered clothes, who curl themselves around earthenware pots, or bathe under cascading waterfalls. The Venus Studio proprietors are sometimes required to photograph urban clients in poses that imitate film stars but the predominantly village resident clients prefer to be photographed full-pose against their painted studio backdrop of Kashmir's Dal Lake or the ruins at Mandu to be found on the facing wall. Both the promotional images and the local use of landscape backdrops, however, testify to the use of photography as a theatrical idiom capable of representing persons with endlessly diverse exteriors, and situated in equally diverse places.

Until the arrival of colour processing after 1987, the main income of Nagda's five main photographic studios came from the photographing and processing of wedding images. Colour films are now despatched to Ujjain for processing and,

* Joanna Woodall (ed.), *Portraiture: Facing the Subject*, Manchester: Manchester University Press, 1997, 131–44.

with the exception of one studio, the photographer's interventions are confined to the production of the negative. The exception is that of the recently established Sagar Studio whose creative proprietor experiments with double exposures on colour negative film. Figure 4.1 is one of these experiments which reproduces imagery he had seen in Hindi films.[1]

A further 500 yards along Jawahar Marg, past the Civil Hospital and Nagda's main liquor shop, is another, older, photographer's premises – those of Suhag Studio. Here Suresh Panjabi, who has been in business for the last two decades, produces from his files a series of exceptional images which depict himself in a variety of double poses. In Fig. 4.2 he represents himself on the left as a politician wearing a hat – a *Gandhi topi* – associated with the Congress Party, which denotes moral probity. On the right of the image he adopts a pose associated with Ramakrishna, a nineteenth-century Calcuttan sage. In other self-portraits the photographer adopts 'poet' and 'dreamer' poses inspired by the mid-1970s Hindi film *Kabhie Kabhie* (Sometimes). The Suhag Studio has been responsible for many of the albums of black and white wedding photographs which many Nagda traders and factory employees (as well as wealthier village landlords) are always pleased to show visitors. These wedding albums compiled by Nagda photographers before

Figure 4.1 Double exposure portrait by Sagar Studios, Nagda, *c*.1993. Original in colour

Figure 4.2 Double exposure portrait by Suhag Studios, Nagda, *c*.1980. Original in colour

the recent switch to colour processing consist of up to 200 or 300 black and white photographs in which great use is made of montage and the collapse of space and time. The bride and groom zoom around India, from the Tower of Victory in Chittaurgarh to the India Gateway in Delhi, to the local Birla Temple.[2]

Like the double or split portraits which are also to be found in these albums (Fig. 4.3), these place a person beyond the space and identity which certain forms of western portraiture, for instance, enforce. They contrast markedly with the dominant Renaissance chronotope within which, to recall Jacques LeGoff's characterization, 'the portrait was triumphant; it was no longer the abstract image of a personage represented by symbols or signs materializing the place and rank assigned him by God, but rather the rendering of an individual captured in time, in a concrete spatial and temporal setting' (LeGoff 1980: 36).

The preference for 'doubles' and 'poses', I will suggest, reflects (and in turn engenders) the lack of a centred, visible 'personality'. In Nagda, photographic portraiture does not necessitate the exploration of the relationship between 'character' and externality within a concrete spatial (and temporal) setting, but is, rather, more likely to represent bodies and faces as infinitely multiple and contingent. Something of the surprising contrast between a dominant western practice and

Figure 4.3 Double printed wedding image by Suhag Studios, 1983

aspects of this local Indian practice is pre-figured in M.K. Gandhi's recollections of his period in England, in the late 1880s, studying at the Bar. Gandhi approached Dadabhai Naoroji[3] for advice as to what an aspirant lawyer should read to succeed in his profession. Naoroji responded that 'a vakil [lawyer] should know human nature. He should be able to read a man's character from his face' (Gandhi 1991: 70). To this end he suggested that Gandhi read Lavater's and Shemmelpennick's books on physiognomy. Gandhi records:

> I was extremely grateful to my venerable friend. In his presence I found all my fear gone, but as soon as I left him I began to worry again. 'To know a man from his face' was the question that haunted me, as I thought of the two books on my way home . . . I read Lavater's book and found it more difficult than Snell's *Equity* and scarcely interesting, I studied Shakespeare's physiognomy, but did not acquire the knack of finding out the Shakespeares walking up and down the streets of London. Lavater's book did not add to my knowledge . . . (Gandhi 1991: 70)

In a similar way, Nagda inhabitants are reluctant to concede that one can 'know a man by his face' but this reflects more than failed casual experiments of the sort undertaken by Gandhi: it is a reflection of a pervasive dualism in which a contingent and mutable external surface is contrasted with a moral character which can be made visible only through action. The customers of Nagda's photographic studios are hugely concerned that their faces should appear free of blemish and shadow, but this is to facilitate a simple recognition of a likeness of the living body of the person depicted rather than any more ambitious revelation of internal character.

David and Judith MacDougall's recent ethnographic film, *Photo Wallahs*, presents a detailed account of practices of disguise and posing in the north Indian hill station Mussoorie, although some of the exegesis suggests differences from Nagda interpretations, In Mussoorie, Indian tourists arrive on Gun Hill by cable car and gaze at the Himalayas, dress as tribals, sheikhs, or parodies of western guitar-strumming hippies. Bodies can be dressed in a limitless range of identities – as Pathan frontiermen, Kashmiri women, gun-wielding dacoits, and village women posing coyly with decorated earthenware pots. One photographer, H.S. Chadha, (MacDougall 1992: 104) offers a client most of the national styles of turban (Rajasthani, Punjabi, Gujarati). This cosmopolitan sartorial excess recalls Appadurai's observation that through its inclusion of regional items much middle-class Indian food simulates a national cohesion (Appadurai 1988), and we may suppose that the clients of Mussoorie photographers aspire to a similar ideal. Edibility and wearability stand as parallel idioms of national integration. It is as if contemporary Indians, rejecting aspects of colonial ethnicisation through physique, costume and other external signifiers, have arrived at a strategy of mutual mimicry, a reciprocal consumability in which – in front of the camera at least – identities are suspended and inverted. Individual images – in which the sitter is clothed in some particular stereotypical garb – can only be understood as part of a much wider repertoire of contingent identities which construct a curiously non-essentialist vision of the nation.

Photographic images of splitting and transformation can be understood against a background of arguments about the existence of 'individuals' within South Asian society. Louis Dumont has long suggested that in India what he terms the basic 'sociological unit' – the Western individual – is entirely lacking (Dumont 1965: 99) since it is subordinated to the interests of the whole. This position has been convincingly challenged by Mines (1988) and McHugh (1989). McKim Marriott has argued that South Asian persons would be better conceptualized as 'dividuals', rather than 'individuals' since they are 'permeable, composite, partly divisible and partly transmissible' (Marriott 1976: 194). Such ethno-sociological 'monist'[4] views also suggest that the moral interior can be read from the exterior. I would not seek to deny the partial validity of such perspectives – for some conditions such as leprosy, or incombustible corpses are recurrently explained by Indians within such a paradigm. But these perspectives may also in part reflect anthropologists' closeness to Brahmanic sources, for ritual specialists are much more likely to claim skills in reading exterior surfaces than lay people. Ascetics are particularly adept at this and most inhabitants in Nagda would accept that an ascetic of any true worth would be able to read the mind and intentions of all those s/he meets. It is this x-ray vision which frequently renders speech unnecessary when devotees encounter ascetics of great power.

But this faith in the visibility of the interior and the readability of exterior signs is not shared by most people in Nagda. The striking dualism they espouse stresses the occlusion of character and the mystery of external surfaces. The majority of people in Nagda are clear that these portraits are only capable of depicting the external characteristics of the sitter, his or her *vyaktitva*. What *vyaktitva* refers to is best denoted by the term 'the signs of being a person'. In Hindi *vyakti* denotes a person, and *vyaktitva* signifies 'person-ness', although it is usually misleadingly translated as 'personality', 'individuality', or 'self.'

In central Indian photographic portraiture the body (*sharir*) is a ground for the physical and visual aspects of *vyaktitva* (complexion, sharpness of features, dress sense – usually ranked in this order of importance).[5] The aspects of *vyaktitva* which a photograph cannot capture might be described as general etiquette – whether a person is a noisy eater, whether they are prone to break wind in public. What photographs are nearly always completely unable to capture, however, is the internal moral character and biography of a sitter, his *charitra*. *Charitra* reveals itself only through actions (*karma*), through past history and future eventualities.

Whereas for the eighteenth-century physiognomist Johann Caspar Lavater, Socrates' immense ugliness posed a problem for the science of physiognomics (he concludes, after numerous pages of discussion, that he is the exception that proves the rule), the inhabitants of Nagda would not be surprised by this disjunction between the external surface and the moral interior. For them it would simply be evidence that one cannot read *charitra* from *vyaktitva*. This is expressed through several idiomatic expressions such as *bahar se kuch aur andar se kuch aur* (one thing outside and something else inside), and *shakal se sidha, lekin kam mem terha* (direct in appearance, but crooked in deeds).

A Jain shop-owner stressed this difference: *vyaktitva* had to do with the exterior surfaces of a person, and was to do with a mixture of the body (*sharir*) and the soul (*atma*) working together. Thus one could at a glance discern the *vyaktitva* of a living person or his/her photo and conclude, for instance, that such-and-such a person did indeed have the *vyaktitva* of, say, a Collector or other important official. In this context *vyaktitva* denotes 'deportment', 'demeanour', or social 'appearance'. Quite distinct from *vyaktitva* there was also the question of *charitra* which was an essentially internal moral quality or character which was only apparent in deeds, that is, in those activities and existential moral decisions which photography – by and large – could not lay bare. So, photographs will reveal only *vyaktitva*, not *charitra*, or they will only reveal the *charitra* of 1 or 2 per cent of photographic subjects. The other 99 per cent will look like the film actors who make themselves look like *mahatma*s and saints, although in fact they are all really *dacoits* (bandits). The quality of dacoitness is a manifestation of *charitra* and can be known only as result of circumstance – by talking and living together.

One might also draw attention here to the observation that in Hindi films disguises are nearly always successful. Usually accomplished by remarkably convincing latex masks worn by the villain, there are no clues of mannerism or voice which serve to reveal the deception. An outstanding example would be the Manmohan Desai film *Mard* (Hero) starring Amitabh Bachchan. At one point in the film both Amitabh, the hero of the title, and his father are convinced by each other's doubles staged by their imperialist enemies and thus goaded to engage in a gladiatorial father–son duel. When Amitabh goes to rescue his father who is chained to a grindstone in a British concentration camp, it is only his perspicacious horse – endowed with extra-natural insight – who senses the deception.

The significance of the distinctions made between *vyaktitva* and *charitra* must also be understood in the context of the widespread acceptance within India of the transience of the body and its status as a contingent receptacle for the soul. Khare has described the expression of this view by Chamar[6] intellectuals in Lucknow, for whom it articulates a political yearning: 'a person's body and caste are his exterior [*upari*] and temporary [*nasvana*] sheaths [*caddar*], while the soul is the imperishable one, which neither dies nor can be higher or lower, but is always equally present in every living being' (Khare 1984: 53). Khare suggests that this Untouchable view overlaps in a limited way with the Kanya-Kubja Brahmans whom he had earlier studied, thus suggesting that this is an ideology that has wide currency, and in Nagda such views are commonly enunciated across a range of different castes.

One important consequence of this conceptualization of the relationship between the evanescent body and the eternal soul is an attenuation of the link between visible and invisible qualities. Because the visible is not deemed to be anchored – in most cases – by an invisible realm of character (for there is usually a disjunction between the two), the external body is thus freed from the constraints with which it is shackled in the western tradition of painted portraiture. What can be captured in a photographic studio is a person's general physiognomy rather than the face as a trace of an interior character. In Nagda, photography works at second remove, with the physiognomy of, say, a deceased relative, making possible the remembrance of an individual which then in turn permits the recollection.[7] There is an attempt to do away with this relay mechanism such that certain physiognomic inflections and nuances are perceived to directly transmit a highly compressed transcript of the sitter's individuality. In Nagda the mode of recognition demands a hieratic clarity – a full-face image with no shadow whose physical recognition is the starting point for the recollection of that individual's life.

Roland Barthes's celebrated search for epiphanic images of his deceased mother recounted in *Camera Lucida*, which has served for some critics as paradigmatic of a western phantasy about photography (e.g. Tagg 1988: 1–3), contrasts sharply with Nagda memorial photography. Barthes describes his quest as not simply for 'just an image', but for a 'just image' (Barthes 1984: 70) which revealed the 'truth of the face [he] had loved' (Barthes 1984: 67). The manner in which Barthes describes the image of his mother which finally achieved what he calls 'the impossible science of the unique being' (Barthes 1984: 71) exemplifies in a hyperbolic form one strand of a western portrait tradition with which Nagda practices have almost nothing in common. Barthes searches through images with a growing discontent:

> I never recognized her except in fragments, which is to say that I missed her being, and that therefore I missed her altogether. It was not she, and yet it was no one else. I would have recognized her among thousands of other women, yet I did not 'find' her. I recognized her differently, not essentially. Photography thereby compelled me to perform a painful labour; straining toward the essence of her identity. (Barthes 1984: 65–6)

Barthes finally finds a photograph of his mother as a child which 'collected all the possible predicates from which my mother's being was constituted' (Barthes 1984: 70) and proceeds to unravel from the compressed cipher of her face her true 'being'. It is not too simplistic to claim that for most of the inhabitants of Nagda known to me, photography is everything which Barthes desires it should *not* be. It is about differential identity rather than 'essential' identity, it is about 'just an image' rather than the obsessive search for a 'just image'.

I would suggest that in Nagda, the studio photographer rarely finds himself asked to picture a subject in the sense of a single being whose totality can be transcribed by a photograph. There are no 'true beings' here waiting to be transcribed. What the Nagda photographer is presented with, by contrast, is a series of bodies, a series of surfaces, objects, planes and angles which can be made to assume different qualities.

These local photographic images need to be understood against a wider backdrop of popular Hindi film and calendar art, within what Appadurai and Breckenridge have termed an 'inter-ocular field' (cf. Appadurai and Breckenridge 1992). There is a recurrent concern within popular Hindi movies with the 'good and bad brother' and collectively these films constitute what is known as the 'lost and found' genre (cf. Thomas 1985) in which two caricatures of *dharma* and *adharma* (duty and the negation of duty) struggle against each other. The clichéd form is a contest between the upright chief policeman of Bombay and a gang boss flooding the city with guns, gold and drugs who finally discover as they lie dying in the arms of their widowed mother that they are in fact brothers separated at birth. As Sudhir Kakar notes: 'In many movies, the "split" in the self is highlighted by the brothers getting separated during childhood. The developmental fate of each remaining unknown to the other, till the final climactic scene in which the confrontation between the brothers takes place and the two selves are finally integrated (Kakar 1981: 7). A variation on this can be seen in the film *Ghazab* (Oppression) released in 1982 and recently reissued. This stars Dharmendra in the classic double role playing a doomed weak brother in the first half of the film and an avenging strong brother in the second half who rights the wrongs done earlier to his sibling. My notes after seeing this film:

> . . . playing Vijay, his muscular Bombay brother, he returns in the second half to settle the score although he is ultimately only able to do this with the help of the deceased brother (now a ghost, referred to as *atma bhai sahab* – 'respected soul brother').

> . . . near the finale Vijay takes his brother's energy (*shakti*) into his own body. Through trick photography we see a literal merging of the two bodies. Vijay's biceps bulge, his shirt rips apart and all the villains are then vanquished . . .

'Splitting', however, is merely one aspect of a much larger spectrum of fragmentation. Threes for instance are almost as popular as doubles: the best filmic example is *Amar, Akbar, Anthony* in which three brothers separated as babies are raised by families of different faiths. In popular calendar art there are numerous images which replicate representations of the *trimurti* (the 'three forms' of Brahma, Vishnu and Shiva). Such arrangements depict the three branches of the armed forces, Kali, Ramakrishna and the Mother, and political trios such as Subhash Chandra Bose, Bhagat Singh and Chandra Shekhar Azad.

Both Bhagat Singh and Chandra Shekhar Azad are famous and celebrated for their terrorist actions in 1931. Today in India they are immensely popular figures and their image together with that of Subhas Chandra Bose (whose Indian National Army fought with the Japanese to rid India of its British colonizers) can be seen far more often than that of western heroes such as Gandhi. Chandra Shekhar Azad was a Brahmin, Bhagat Singh was raised a Sikh. For many Britons in India conditioned by early anthropology's anthropometric obsessions, this would have suggested a fundamental difference, a separation that could be testified to by sartorial, physiognomic and physiological incompatibility. In the popular Indian representation however (such as Fig. 4.4) they appear identical – Indian brothers distinguished only by the presence of a hat or a wristwatch and sacred thread. Images of Singh and Azad became hugely popular in the 1930s and many proscribed images dating from this period can be seen in the Proscribed Indian Books collection in the India Office, London. Figure 4.4 is a calendar image painted by

Figure 4.4 Bhagat Singh and Chandra Shekhar Azad. Colour calendar image by H.R. Raja, *c*.1980

H.R. Raja, a Muslim artist based in Meerut who appears to have first developed this physiological similitude and to have elaborated the theme in many hundreds of paintings in the last two decades.

Raja's image suggests an oppositional practice, a political critique of the divisiveness of British policy and much of its imperial science, It also, however, demonstrates how the face in contemporary Indian representations can be erased of its physiognomy, its epiphanic qualities effaced in favour of more compelling arguments. Inasmuch as the similarity of action and political identity (they were both fighters for the freedom of India) is signified by their identical bodies, this supports a monist interpretation. More persuasively, however, it might be argued that sameness can be exported between different bodies and faces because there is no internalized and coherent subject of portraiture.

That photographic portraiture in central India is not centred on a physiognomic personality is perhaps also demonstrated by certain practices associated with memorial portraiture. A long-term acquaintance of mine, Hira, a scheduled caste Chamar, was recently killed by a train and his mother took some group photographs which included Hira to a Nagda Studio where his figure was isolated and emphasized by the application of concentric rings of blue, red and yellow, which are often used as divine haloes in chromolithographs of deities. These rings were painted on before the image was rephotographed with an enlarging lens (Fig. 4.5). This is one example of the manipulation of the photographic image available to very poor bereaved villagers. In Nagda, by contrast, wealthier bereaved families have access to an altogether more complex process.

Many of the painted memorial photographs to be seen in shops and homes in Nagda are the work of Nanda Kishor Joshi (Fig. 4.6), an itinerant Brahman artist from Beawar, Rajasthan who visits Nagda several times yearly in search of commissions. The majority of these are for memorial 'photos' of deceased family members. Nanda Kishor Joshi's artistry is concerned with perfecting the past, rendering the transient flux recorded in photographic emulsion into more permanent, more true forms. Everything about Nadga photographic practices suggests that no value is placed on photography's documentary ability to record the random and inconsequential. On the contrary, photography is prized for its ability to record idealized staged events characterized by a theatrical preparedness and symmetry. Nanda Kishor Joshi's economic niche is to be found precisely in the space between photography's indexical randomness and Nagda people's demands for images constructed according to a significantly different aesthetic.

Figure 4.5 Colour print of Hira overpainted by an unknown Nagda photographic studio

Figure 4.6 Nanda Kishor Joshi holding a photograph which will form the basis of a memorial portrait

When clients give him a photograph (he prefers black and white) of the deceased, he also completes a form of 'particulars' (Fig. 4.7). In this is recorded the desired colours of (in this order): hair, head-dress, shirt, coat, jacket, sari, blouse, ornaments, eyes, *bindi* or *tilak*,[8] trousers, petticoat. Following this there is a section for 'special instructions' and for the name which is to be written on the portrait. The form finally notes that these images are best seen from approximately eight feet away.

The original photograph – often extremely small – is scaled up by Nanda Kishor Joshi to the required size of the new portrait (Fig. 4.8) and the 'special instructions' are incorporated in the final work which, once mounted and framed behind glass, costs in the region of Rs. 300–450.[9] Nanda Kishor[10] stresses the flexibility which his style of world-making offers:

Figure 4.7 Order form for 'Hemu Art Center' and original photograph given to Nanda Kishor Joshi

whatever a person wants can be put in the photo . . . make the clothes this colour, even put new clothes in – if no *kurta* is worn, no coat is worn, then a coat can be given, if there is no *topi* [a cap, in this context mostly worn by Jains], a *topi* can be provided, a *pagri* [turban] made – Marwari, Rajasthani, Panjabi, Gujarati, Ratlami [styles of *pagri*] – whatever design a person wants can be provided.

Nanda Kishor shows me his current commissions. He points out various additions in his coloured portraits: 'a Jodhpuri *pagri* . . . *kangressi topi* [the cap associated with Congress Party politicians] . . . a gold chain around a Jain's neck . . .', complete transformations from *kurta* in the original photo to a Jodhpuri coat covering his throat. Others asked for *tilaks* (a ritual mark in the centre of the forehead) or they might say 'there's a *pagri*, give me a *kangressi topi* . . . give me a white shirt . . . give me a moustache'. One man had been killed by a train and his father wanted a photo but with more hair and the skin colour lightened; another wanted a red *pagri*; another wanted a face 'shaved'.

Sri – of Nagda had given him a coloured photo: 'He's alive [*zinda hai!!*], not deceased.'

He wants his *pagri* the same colour as in the photo, but needs shaving and he wants a closed Jodhpuri coat. No. 867 a Porwal [middle ranking trader caste] . . . his *topi* is too high on his head, it will be lowered a little. The collar on his *kurta* is open a little too

Figure 4.8 Completed memorial image by Nanda Kishor Joshi together with its original photographic referent

much . . . it will be closed a little. One woman, with severe *khol* [vitiligo – lightening of the skin] . . . the original colour will be restored and a *bor* [a brass ornament] placed in her hair . . . both her eyes will be straightened . . . she's old and one eye is going over there and one over there.

The transformation of the past is perhaps starkest where Nanda Kishor has to rely on post-mortem photography. Photographs are often taken by relatives and studios after death when studies from life exist, but they are given to Nanda Kishor only when they are the sole image available. In such cases the physical manifestations of death can be turned back, mouths can be closed, and eyes opened as the *yadgari* (memento) takes shape.

The contrasts with Barthes's search for the 'just image' of his mother are illuminating. Nanda Kishor's images lack the compressed epiphanal transcript that Barthes desperately searched for, and his clients do not yearn for an 'impossible science of the unique being': their quest is for the 'merely analogical, provoking only . . . identity' (Barthes 1984: 70–1). This must in part be because of their function – they are public icons, often displayed in shop fronts and worshipped monthly on the *tithi* (lunar day) of the ancestor's death and annually during a more elaborate ritual procedure in *pitra paksya* (ancestors' fortnight). But they also have to do the work of remembrance and will often be the only images of the deceased which are carefully kept. Similarly, Nanda Kishor Joshi's painted photographs suggest a striking answer to the question posed in the eighteenth century by Johann Caspar Lavater: 'where is the art, where the dissimulation, that can make the blue eye brown, the grey one black, or if it be flat, give it rotundity?' (Lavater n.d.: 84)

Coomaraswamy would have described these images as 'effigies' rather than 'portraits' (Coomaraswamy 1977: 89). This is not merely because he argued that 'portraits' are 'likenesses of a person still living', but also because, as he percept-ively noted, 'portraiture in the accepted sense is history'. This recalls LeGoff's characterization of the true subject of portraiture as an individual configured in a 'concrete spatial and temporal setting' (LeGoff 1980: 36). The dominant tradition of western portraiture in oils has given expression to such a chronotope framed by a perspectival window and much western photography has in many ways repro-duced the conventions established within this earlier tradition of painting.

Portrait photography in Nagda is not imprisoned by such a representational chronotope but, rather, has its own quite distinct constraints. Photography, rather than providing a perspectival window in which to ground persons, is, in this Indian context, a means through which individuals can be displaced. Clifford Geertz once highlighted a crucial facet of 'western' conceptualizations of the person as its 'organiz[ation] into a distinctive whole and set[ting] contrastively both against other such wholes and against its social and natural background' (Geertz 1984: 126). Nagda photography displays little evidence of this for, as we have seen, photography is used as a means to set aspects of a person against aspects of that same person. Rather than seeking to provide a ground in which solidity can be conjured up through oppositions with similarly grounded individuals, photography appears as a tactic for self-enquiry. The space in which this occurs is not the tech-nological picture plane that some photography shares with western oil painting, but the flat space of the photographic positive in which doubles or triples of the same displaced individual can be conjured.

Notes

1. Eyes, both divine and human, are frequently the subject of complex sequences in popular cinema and reflect the centrality of vision in Hindu ritual encounters. The commonly used term *darshan* translates as 'seeing and being seen by the divine'. Cf. Eck (1985).
2. For an example see Pinney 1992: 42.
3. Naoroji represented Finsbury in the House of Commons as the first Indian Member of Parliament and was also President of the Indian National Congress.

4. Here, monism denotes a denial of the duality of mind and matter and the proposal that physical and psychical phenomena are interpenetrating.
5. As we saw earlier with reference to a 'Collector', these features may also be indicative of social and professional status.
6. A scheduled caste (formerly 'Untouchable') whose traditional occupation is leather-tanning.
7. I mean here the dominant tradition that presents the face as a transcendent sign (cf. Koerner, 1986 but I also acknowledge the presence of other western traditions (e.g. the 'swagger' portrait) which have greater similarities with the twentieth-century central Indian images discussed here.
8. Marks on the forehead.
9. Approximately £6–9 at 1995 exchange rates. Though a small sum for many prosperous Nagda traders, it might be contrasted with agricultural labour rates of around Rs. 20–25 per day.
10. The following are based on my translations of extensive taped interviews in 1991 and 1992.

References

Appadurai, A. (1988), 'How to Make a National Cuisine', *Comparative Studies in Society and History* 30: 3–24.

Appadurai, A. and Breckenridge, C. (1992), 'Museums are Good to Think: Heritage on View in India', in I. Karp et al. (eds), *Museums and Communications: The Politics of Public Culture*, Washington DC: Smithsonian Institution Press, 35–55.

Barthes, R. (1984), *Camera Lucida: Reflections on Photography*, London: Fontana.

Coomaraswamy, A. (1977), 'The Part of Art in Indian Life', in R. Lipsey (ed.), *Coomaraswamy, Selected Papers I*, Princeton: Princeton University Press, 71–100.

Dumont, L. (1965), 'The Functional Equivalents of the Individual in Caste Society', *Contributions to Indian Sociology* 7:

Eck, D. (1985), *Darsan: Seeing the Divine Image in India*, Chambersburg: Anima Books.

Gandhi, M.K. (1991), *An Autobiography*, Ahmedabad: Navajivan Publishing House.

Geertz, C. (1984), 'From the Native's Point of View: On the Nature of Anthropological Understanding', in R.A. Shweder and R.A. LeVine (eds), *Culture Theory: Essays on Mind, Self, and Emotion*, Cambridge: Cambridge University Press,

Kakar, S. (1984), 'The Ties That Bind: Family Relationships in the Mythology of Hindi Cinema', *India International Centre Quarterly* 8(1):

Khare, R.S. (1984), *The Untouchable as Himself: Ideology, Identity, and Pragmatism among the Lucknow Chamars*, Cambridge: Cambridge University Press.

Koerner, J. (1986), 'Rembrandt and the Epiphany of the Face', *Res: Anthropology and Aesthetics* 12: 5–32.

Lavater, J.C. (n.d.), *Essays on Physiognomy*, London: William Tegg.

LeGoff, J. (1980), *Time, Work and Culture in the Middle Ages*, Chicago: University of Chicago Press.

MacDougall, D. (1992), 'Photo Hierarchicus: Signs and Mirrors in Indian Photography', *Visual Anthropology* 5:

Marriott, M. (1976), 'Interpreting Indian Society: A Monistic Alternative to Dumont's Dualism', *Journal of Asian Studies* 36 (3):

McHugh, E.L. (1989), 'Concepts of the Person among the Gurungs of Nepal', *American Ethnologist* 16: 75–86.

Mines, M. (1988), 'Conceptualizing the Person: Hierarchical Society and Individual Autonomy in India', *American Anthropologist* 90 (3): 568–79.

Panofsky, E. (1991), *Perspective as Symbolic Form*, New York: Zone Books.

Pinney, C. (1992), 'The Lexical Spaces of Eye-Spy', in P.I. Crawford and D. Turton (eds), *Film as Ethnography*, Manchester: Manchester University Press.

Tagg, J. (1988), *The Burden of Representation: Essays on Photographies and Histories*, London: Macmillan.

Thomas, R. (1985), 'Indian Cinema: Pleasures and Popularity', *Screen* 26 (3–4): 116–31.

–5–

Heritage and Cultural Property
Michael Rowlands

Introduction: Heritage, Politics and Memory

The purpose of this chapter is to outline some of the implications of the fact that we live in an era of unprecedented concern with preserving and restoring the past. What is striking about the present is how terms like heritage, tradition and cultural memory have slipped into everyday speech as unproblematic ways of accounting for the right of people to feel that they belong somewhere. A number of factors account for this. Our attention is drawn to common concerns with creating identities, with the need to establish what it means to belong and the fact that this is no longer debated as a private matter according to personal disposition (if it ever really was) but has become a question involving the definition of the public sphere and civic consciousness.

In an influential article on the rise of cultural fundamentalism in Europe, Verena Stolcke claims that "rather than asserting different endowments of human races, contemporary cultural fundamentalism . . . emphasises differences of cultural heritage and their incommensarability' (Stolcke 1995: 4). She argues that a new politics of recognition has arisen since the early 1990s that links a politics of identity with a need to belong. One of the more challenging implications of recent debates on globalization has been to show that an obsession with belonging is the flip side of increasing global flows of recent times (Friedman 1992, Harrison 1999, Meyer and Geschiere 1999) and as Appadurai (1996) showed, the 'production of locality' is crucial to this redefinition of belonging. It appears that the maintenance of borders and boundaries remain crucial under conditions of globalization, even though they may be in constant flux. This corresponds with a growing resistance to those perceived to be outsiders (anthropologists, tourists, patent agents or migrants) who it is feared might appropriate the cultural knowledge, identities and economic resources of indigenous minority peoples (Brown 1998). The right to an identity has now become the basis for making claims to intellectual and cultural property which have diversified into a spectacular range of knowledges including practices as diverse as music and medical pharmacology (Coombe 1998, Strathern 1999). Finally the growth of neoliberal demands for authoritarian states to democratize also appear to have contributed to intensifying a politics of belonging. New forms of social exclusion which cite the rights of those claiming indigenous origin over those of migrants and strangers, turn citizenship into a hotly debated issue whilst at the same time justifying policies of ethnic cleansing and the destruction of cultural heritage (Geschiere and Nyamnjoh 2000, Layton, Thomas and Stone 2001).

These trends have encouraged an increasing concern with the identification and protection of cultural heritage. Lowenthal (1996) dates the rise of a heritage and museum boom from about 1980 which he associates with the appearance of new Right Wing agendas both in America and Europe. He claims they share similar roots in a new preoccupation with ensuring the transmission of cultural legacies from past to future generations. The destruction of cultural heritage, for example accompanying ethnic cleansing policies in Bosnia, reminds us of the passions and violence that can now surround the right to possess a culture (Layton, Thomas and Stone 2001). Heritage is also big business, whether measured by catalogue prices at Sotheby's or as an attraction for tourists or as a billion dollar conservation industry. What Derrida (1996) calls 'archive fever' now suffuses a global culture obsessed with accumulating evidence of cultural difference. The UNESCO World Heritage Centre has so far legislated 552 sites for protection worldwide which also justifies state legislation and new definitions of civic consciousness in the guise of protecting national heritage. In England alone there are more than 657, 000 registered archaeological sites with the number having increased by 117 per cent since 1983 (Holtdorf 2001: 288). The fear should not be that we will have too little of our past left in future but that more and more of our world will be recorded as some sort of historical or natural site worthy of preservation. What is the perceived need behind these political and cultural agendas that justify simultaneously the dominance of the past, the commodification of the past and a search for authenticity in the formation of cultural identities?

The invention of 'cultural heritage' was bound to powerful mythologies which seek to reclaim and possess lost pasts, return to imagined homelands and redeem the wisdom of ancient Golden agés (Lowenthal 1985). In terms of the experience of a particular Euro-American view of modernity, heritage meets the need to salvage an essential, authentic sense of 'self' from the debris of modern estrangement. What might be called the 'seduction of heritage' lies in its promise to redeem a past and in so doing to be the means to revive a sense of authentic being in the contemporary world (Benjamin 1977). The potency of heritage revivalism continues to have effect as a means to 'cure' postmodern identity crises and to counteract late modernist experiences of rootlessness, rupture and displacement (Jameson 1991). Lowenthal (1996: 6) tells us that beleaguered by a sense of loss and change, we keep our bearings only by clinging to the past as remnants of stability. Heritage is infused by a sense of melancholia and grief for lost objects and lost sense of identity. As a western phenomenon it is part and parcel of the rise of capitalism and fears of the erosive powers of market economies. Since the eighteenth century in Europe, collections of cultural artefacts and works of art have been associated with the need to preserve 'things of value' for posterity as a critique of the dissolving of social relations and identities by the rise of commerce. The heritage industry has also ironically been accused of commodifying 'authentic pasts' for purposes of tourism, advertising in order to sell souvenirs and tourist art (Phillips and Steiner 1999). Proponents of the cultural heritage industry are accused of wanting to commodify the past and turn it into a tourist kitsch or a Disney-fied theme park version of history in place of the real thing (Samuel 1996: 259). As a critique it can be seen as the latest manifestation of what Debord termed the trend for the contemplation of things to be replaced by spectacle, outrage and distraction (Debord 1968).

Others have also argued that the rise of heritage in the 1990s has become the single most important vehicle for defining identity. The philosopher Charles Taylor (1994) has accounted for this new obsession as part of a more general trend – the right to recognition. He argues in the context of a debate on multiculturalism, that a politics of recognition has supplanted a politics of redistribution as part of a trend towards the equalization of rights and entitlements. Respect, as a precondition of recognition, is what we all struggle for in our intimate relations with others. However what has emerged in the guise of neoliberalism is a larger role for a politics of recognition in the public sphere (Taylor 1994: 37). Recognition has now become a precondition for rights and claims to entitlements because it is compatible with a more market oriented view of identity. A politics of redistribution which has characterized the welfare economies of the post Second World War era, considered injustice and inequalities could only be remedied through economic means. Hence it was argued that cultural differences were of minor importance relegated to the private sphere and the need for these would wither away as affirmative redistribution tackled the conditions of socio-economic inequality. In the 1990s the implications of neoliberal policies, structural adjustment programmes, the end of 'cold war' conflicts, have all conspired to produce a new consensus that the allocation of resources will in future depend on the progress of a politics of rights. Yet the assertion of rights (human and cultural) implies a more basic right to recognition. But recognition of what?

Cultural Rights and Wrongs

Two rather novel assertions about the right to possess a culture can be distinguished in recent international conventions on cultural rights. One is the value of self-awareness about belonging to a particular culture and the right to preserve it. Cultural practices which were 'taken for granted' now become self-conscious acts and form the bases for political action. This can of course be symptomatic of a certain crisis of confidence about possessing a culture and the need to objectify what is left. In other words people not only live 'their culture' but will now reflect upon, evaluate, discuss, modify and dispute it, and this takes increasingly politicized forms (cf. Harrison 2000: 663). The second is that in possessing a culture, a legal discourse of rights has created an ambiguity precisely through the legal definition of culture as property (Brown 1998, Strathern 1996).

In a western-derived legal discourse of rights, authorship designates both a legal status of ownership, a mode of aesthetic production and a form of moral subjectivity. In particular, the relationship between the category of the author in copyright law and various ideas about the aesthetic personality of the author as unique and original, raises ethical questions about how to define the status of unique identities against the homogenizing tendencies of a transcultural language of rights. A romantic emphasis on originality and creative genius had previously functioned to marginalize or deny the creativity of others, privilege notions of 'high culture' and justify the right to assimilate the creativity of others. Objectified notions of culture and the right to possess a culture would appear therefore to be an ethical precondition of a politics of recognition and to collude in justifying acts of cultural destruction, if it means that this allows one the right to assert a unique identity. Worries about overconfident assertions of cultural rights has generated a dialogue about how to qualify a language of rights with the meaning of culture. Rights to intellectual or cultural property have been based on claims to preserve 'a culture', to have access to a 'lost culture' and the need to sustain cultural diversity as an analogue to biodiversity (Strathern 1996). Since the late 1990s there has been a full-scale retreat from appearing to impose universal definitions of cultural rights on the bearers of 'traditional cultures'. Objections by indigenous rights movements to their unique identities being reduced to a generalized language of cultural rights has prompted a more relatavist reponse in heritage legislation (e.g. Prott 1998). A politics of recognition has also been empowered by cultural heritage precisely because there is something about the materiality of the latter, about its escape from the ambiguities of language and writing, that makes it peculiarly responsive to claims of authenticity and possession of unique identities.

We are encouraged to think of heritage as a discursive practice, as a way that a group slowly constructs a collective memory for itself by telling stories about itself. These stories constitutes a tradition which in the structure of a narrative constitute a relation between a past, a community and an identity to define the right of a group to exist. A tradition is therefore never neutral but is created by an aesthetic relationship between objects, memories and stories which can transmit to future generations a sense of dignity, self-respect and a right to have a future. If heritage implies a common possession, as David Scott argues: 'it does not presuppose uniformity or a con-sensus. |Rather it depends upon a play of conflict and contention. It is a space of dispute as much as of consensus, of discord as much as accord' (Scott 1999).

Heritage as discourse implies on the one hand closure around a single legitimate narrative and on the other acceptance that more than one story can be told. This gives the idea of 'heritage' a curious inflection. Whatever is described or identified as heritage is only valuable as long as an authentic sense of past exists to identify it as such. There must therefore already be an accompanying story, of the nation or a community for example, whose terms we already know and for which 'heritage' provides a material objectification or a collective representation of a particular version of tradition (cf. Hall 2000). 'Things of value' have to be preserved in order to justify and legitimize these accounts. And as Hall argues (2000: 4), since the eighteenth century, these have been increasingly the monopoly of the state and its control of public education and the desire to manage 'civic consciousness'. Collections of cultural artefacts, the preservation of monuments and in general the inculcation of civic pride in an historical landscape has been part of associating a 'beyond questioning' sense of belonging with the identity of the state. Like personal memory, cultural memory is highly selective; in Hall's terms, it highlights and foregrounds whilst at the same time it silences and disavows, eliding those episodes that might form the opportunity for alternative narratives (Hall 2001: 5). The theme that heritage is a site of contestation is recognized by archaeologists debating the meanings attached to sites and monuments by new social movements (e.g. Bender 1998). This implies that 'heritage' can also be the source of alternative narratives because the materiality of the past need not accompany an existing story but might lie uneasily and be a source of contradiction with it. Responding to the question who the past is for by the answer 'those who belong to it' only raises the additional question therefore of who should have the power to define heritage and therefore the culture of belonging and who this includes and excludes. A point that has hastened a crisis in the authority of museums and public cultural institutions.

Heritage and a Critical Museology

A 'revolution' has occurred in museum studies in the last two decades as a consequence of a more general critique of the Enlightenment ideal of dispassionate knowledge. In the past national heritage was embodied in the national museum whose collections and exhibitions formed a living encyclopaedia of universal knowledge. Going back to the ideal of the classical *museon*, national museums such as the Louvre, the British Museum or the Metropolitan Museum were established to show how the universal development of culture could be linked to the development of national history and how the latter could be equated with the very notion of civilization itself. The West, as the outcome of the development of civilization, developed with a particular twist as the principal heir of the Enlightenment rediscovery of rationality and science.

The construction of grand narratives was already the subject of internal critiques ranging from Said and Bernal to Lyotard and other postmodern philosophers. But as a consequence of a series of controversial exhibitions such as the 'Magiciens de la Terre' at the Pompidou Centre or the ' te maori' exhibition at the Metropolitan Museum in New York, the exhibiting and display of 'other cultures', regardless of the best intentions of the curators, became the subject of controversy (Clifford 1988). The question raised is who has the right to represent the cultures of others and if so does this scepticism of motives justify cultural relativism. The impact of a growing reflexivity about the constructed and therefore contestable authority of ethnographic museums has brought sharply into focus not only the question of representing others but also the holding on to collections that by and large could be argued to be the product of colonialism and exploitation and in some cases of looting and plunder (Simpson 1996). As far as the question of representation is concerned it is unlikely now that any major exhibition based on ethnographic collections would go ahead without detailed consultation with indigenous curators and a commitment to developing a joint strategy for the conservation and research of collections. Several of the major national museums have redesigned their brand logos to stress their role in the comparative understanding of world cultures. The recognition that authenticity may be a difficult quality to espouse in the face of hundreds of years of colonial/postcolonial interaction and dependence also encourages debate that hybrid identities may be a more realistic avenue to reclaiming the past in multicultural settings.

But the problems encountered by indigenous groups claiming minority status in situations where the dominant culture are descendants of white colonial settlers has established quite the opposite trajectory. Whilst varying in their histories of colonialism, indigenous communities share similar experiences in achieving recognition only through direct assertion of claims for absolute autonomy. Participating in a nation-building rhetoric of forging diversity in unity is a tricky affair when you have few resources to maintain an independent identity. Indigenous rights depend upon recognition of prior origin and original ownership of land, that have subsequently been usurped and appropriated by later colonial settlers. Heritage is reconstituted as cultural property as long as it helps legitimize claims in the public sphere dominated by the colonial or postcolonial state. Indigenism is perhaps one of the fastest growing strategies for re-imagining local identities in a context where the production of locality is based on claims to origins. And in a paradoxical manner, it is associated with an enormous creative explosion in the arts that ironically is concerned not with indigenism as such but with breaking out of an ethnic closure of tourists' arts, world music and ethnic products into the mainstream of cultural production. Such transcultural or globalizing 'origins' are complex in their motives and in the diversity created within a transnational 'civilizational logic'. To discuss Hindu film or music for example is an immensely more complex phenomenon today not only because of the forms of diaspora created through migration but also because of the influence it wields in more hybrid cultural terms (Dwyer and Pinney 2001). The cultural repertoires of heritage, constantly worked and reworked by each generation, now bypass the boundaries of state and nation and create scales of public culture and identity that are world systems of practice and belief (Appadurai 1996).

Cultural Memory, Loss and Redemption

Monuments, buildings and ruins have the obvious merit of time literally being inscribed on the surface of things. What Nora 1989 has called places of memory share in common this 'dream-like' quality which convinces us that reality also retains an iridescent quality of an enchanted, timeless world regardless of all the evidence to the contrary. Duration, in which time is literally inscribed as age, preserves both personal and collective memories in cultural form and as aides-memoires. Such props of memory prompt recall not as a memory trace but as an internal dialogue or representation that disturbs the turning of a lost object into a past that can then be forgotten. Objects also act as an aide-memoire for an external conversation or a narrative about a person or event that is in danger of being forgotten. The double aspect of memory-work – remembering and forgetting – has unpredictable consequences as far as which object will act as a memory prop (Forty and Küchler 1998).

Heritage also implies a threat of loss and the need to preserve or conserve against an inevitable sense of deprivation. Recognition equally implies a past trauma around a loss or sense of deprivation that can only be healed though the drive to closure and forgetting. Heritage is therefore also a metaphor for the demand for justice and the resolution of a grief that remains a wound. There are many reasons why forgetting should be the desired outcome but amnesia as repression implies only temporary resolution whilst for Freud real forgetting could only take place after adequate remembering. A compelling concern for heritage studies at the present time therefore is the right to transmit a sense of loss and unresolved grievance to the future and how the destruction of objects as memory props endanger the prospect of recognition. Freud termed all such situations as unfinished emotional attachments to lost objects which can neither be internalized or reattached to other objects. Instead of being able to turn such lost objects into a past that can be forgotten, a sense of loss is retained as a grievance to be idealized, projected and defended (Feuchtwang 2000, Rowlands 1998).

The idea that heritage is therefore selective and is concerned as much with the ability to forget as to remember, motivates the concern by Hall, Lowenthal and others that tradition is (in contrast to history) always mobilized around the issue of cultural amnesia and original acts of violence. When in 1882 Ernest Renan (1990: 11) made his famous remark that forgetting was a necessary condition for the creation of the nation, he also noted that 'every French citizen has to have forgotten the massacre of St Bartholomew' (1572 of the Huguenots). Cultural amnesia is motivated to forget the traumatic events that endanger a temporary sense of unity and totality (e.g. Hall 2001). The idea that subjects, whether nations or persons, emerge from and are sustained by amnesia is then confirmed in their own self-recognition, first as victim and then as perpetrator. This no doubt is why most acts of violent origins are successfully forgotten or converted through the language of sacrifice into past voluntary and creative acts of renewal. Freud's (1957) concern with how mourning can be closed and the likely impossibility of this being fully achieved is balanced by the strength of the drive to closure and the search for justice.

The question begged is whether 'heritage cultures' and their representation in museum exhibitions, theme parks and in advertising are conservative and primarily concerned with the anodyne, the romantic and with maintaining a sense of closure or whether they can form a practical critique that opens the opportunity of cultural transgression. Critique also implies a seeing beyond the foundational and the implicit to a future where 'things might be otherwise'. The search for knowledge through formal structures and universal values may be a reactionary path serving to close off investigation or it can be a critical path that opens up alternatives which provide, through recognition, redress to past injustices and sufferings. Ideology triumphs for those who are able to exclude certain forms of memory work or deny some the right to transform experiences of trauma into narratives of belonging and regeneration. There are clear signs that heritage and a critical museology are having to respond to these major challenges. The notion that museums as national institutions need only portray dominant values (of the nation, a dominant class, etc) has been undermined by the explosion of interest in a 'history from below'. Museum exhibitions on 'the slave trade', on labour history, on minorities in multicultural Britain for example have also started to undermine the racial stereotypes of a political history that sees the constitution of the nation separate from empire. Critiques of the Enlightenment ideal of dispassionate knowledge coupled with a rising cultural relativism have led to the undermining of Western or Eurocentric grand narratives about the cultural heritage of others. Demands from indigenous, minority and suppressed cultures to repossess their heritage has wielded a critical edge, confronting the heritage culture with its own complicity in the often violent appropriation of land, artefacts (including secret sacred material), human remains and in the scientific, cultural and intellectual colonization of other cultures. (Greenfield 1996, Layton 1990, Simpson 1996). What has emerged is a significant change in our awareness of the centrality of culture as something owned and its relation to identity and how to conceptualize it. Newer, more active and inclusive expressions of 'heritage' and museology need to be pursued.

Anthropologizing the Heritage Debate

Much faith is put at present in the transforming qualities of both museum and heritage cultures. By being able to assert the right to have a culture and identity, re-engaging with the past is claimed to be not just a matter of re-call but to have a more curative role. Debates over repatriation of cultural property, or the return of human remains to their living descendants for reburial, or demands that majority, mainstream versions of heritage should recognize the implicit presence of others in common demands for restitution and redress through the recognition of past injustices. Fundamentally therefore these identities are preserved because heritage is centred around other discourses which make up 'the politics of recognition' i.e. desires for and rights to 'dignity', respect and 'social justice' and demands for alternative mechanisms for the allocation of resources in society.

This implies that academic subjects (including in particular anthropology, archaeology and museum studies) cannot really grapple with these issues at the present time because they lack a sufficiently subtle and nuanced approach to cultural heritage. In some respects this mirrors anthropology's rather late appraisal of the significance of landscape as a means of relating issues over land conflicts with claims legitimized by rights to sacred sites and notions of environment that have no objective qualities if evaluated in terms of individualized notions of property (cf. Bender in this volume). The pressures for the academy to accept the significance of indigenous ethnography as different in truth claim from western notions have taken time to be recognized and not without considerable resistance. Moreover it is clear that anthropologists who do research in the Americas or Australia have generally been both more sensitive to questions of difference which are not binary (either/or) and therefore more willing to recognize the implications for their discipline of sharing complicated colonial pasts with their subjects.

Heritage, at first a discourse of Euro-American origin, now a feature of globalization, occupies a number of distinct registers. For example heritage in Britain, evoking the cultural complexity of multiculturalism, demands recognition of the facts of a history of empire in British identity, (e.g. of an African presence in Britain since the sixteenth century) and a demand for justice, i.e. to be treated equally. Heritage in Europe is also the heritage of the diaspora, of the global experience of migration and the desire for a homeland. Finally there is the question of those 'traditions' which on the one hand make claims to possess distinct 'archives' of 'civilizational proportions' and on the other to provide the stimulus for a rich profusion of contemporary hybrid or crossover cultural forms. All of these represent some of the most important cultural developments of our time and explain why the question of heritage and the forms by which it is archived and represented is of such timely and critical importance at the moment. Which is also a timely reminder that heritage whilst ostensibly about the past, is always about the future. About having a future and the recognition that those excluded or marginalized from this discourse are also being denied one.

References

Appadurai, A. (1996), *Modernity at Large*, Chicago: University of Chicago Press.

Bender, B. (1998), *Stonehenge*, Oxford: Berg.

Benjamin, W. (1977), 'Theses on History', in *Illuminations*, London: Fontana.

Brown, M. (1998), 'Can Culture be Copyrighted?' *Cultural Anthropology* 39: 193–222.

Clifford, J. (1988), *The Predicament of Culture*, Cambridge: Harvard University Press.

Coombe, R. (1998), *The Cultural Life of Intellectual Properties*, Durham, NC: Duke University Press.

Debord, F. (1968), *Society of the Spectacle*, London: Routledge.

Derrida, J. (1996), *Archive Fever*, Chicago: University of Chicago Press.

Dwyer, R. and Pinney, C. (2001), *Pleasure and the Nation.* New Delhi: Oxford University Press.

Feuchtwang, S (2000), 'Reinscriptions: Commemoration, Restoration and the interpersonal transmission of histories and memories', in S. Radstone (ed.), *Memory and Methodology*, Oxford: Berg.

Forty, A. and Küchler, S. (1998), *The Art of Forgetting*, Oxford: Berg.

Freud, S. (1957), 'Mourning and Melancholia, in *The Works of Sigmund* Freud, Standard edition vol 14, E. Jones (tr.), London: The Hogarth Press.

Friedman, J. (1992), 'The Past in the Future: History and the Politics of Identity', *American Anthropologist* 94(4): 837–59

Geschiere P. and Nyamnjoh, F. (2000), 'Capitalism and Autocthony: the seesaw of mobility and belonging', *Public Culture* 12/2: 423–52

Greenfield, J. (1996), *The Return of Cultural Treasures*, Cambridge: Cambridge University Press.

Harrison, S. (1999), 'Cultural Boundaries', *Anthropology Today* 15: 10–13.

Harrison, S. (2000), 'From Prestige Goods to Legacies: Property and the Object-ification of Culture in Melanesia', *Comparative Studies in Society and History*, 42: 662–79.

Hall, M. (2001), 'District Six Museum', in B. Bender (ed.), *Exile*, Oxford: Berg.

Hall, S. (2000), 'Whose heritage?', *Third Text* 49: 3–13.

Holtdorf, C. (2001), 'Heritage as a Renewable Resource', in Layton, Thomas and Stone (eds), *The Destruction and Conservation of Cultural Heritage*, London: Routledge

Jameson, F. (1991), *Postmodernism, or the Cultural Logic of Late Capitalism*, London: Verso.

Layton, R. (1990), *The Politics of the Past*, London: Unwin Hyman

Layton, R., Thomas, J. and Stone, P. (2001), *The Destruction and Conservation of Cultural Heritage*, London: Routledge

Lowenthal, D. (1985), *The Past is a Foreign Country*, Cambridge: Cambridge University Press.

Lowenthal, D. (1996), *The Heritage Crusade*, London : Viking.

Meyer, B. and Geschiere, P. (1999), *Globalisation and Identity*, Oxford: Blackwell.

Nora, P. (1989), 'Between Memory and History', *Representations* 26: 7–25.

Phillips, R. and Steiner, C. (1999), *Unpacking Culture*, Berkeley: University of California Press.

Prott, L. (1998), 'Understanding one Another on Cultural Rights', in H. Niec, (ed.), *Cultural* Rights and Wrongs*, Paris: UNESCO.*

Renan, E. (1990), 'What is a Nation?', in H. Bhabha (ed), *Nation and Narration*, London: Routledge.

Rowlands, M. (1998), 'Remembering to Forget: Sublimation in War Memorials', in A. Forty, and A. Küchler (eds), T*he Art of Remembering*, Oxford: Berg.

Samuel, R. (1996), *Theatres of Memory*, Oxford: Blackwell.

Scott, D. (1999), *Refashioning Futures*, Princeton, NJ: Princeton University Press.

Simpson, M. (1996), *Representations*, London: Routledge.

Stolcke, V. (1995), 'Talking Culture: New Boundaries, New Rhetorics of Exclusion in Europe', *Current Anthropology* 36: 1–26.

Strathern, M. (1999), *Property, Substance and Effect*, London: Athlone.

Taylor, C. (1994), *Multiculturalism*,

The Power of Origins: Questions of Cultural Rights*
Michael Rowlands

Introduction

In 1929, a series of public lectures was delivered at University College London on the subject of the 'Englishman'. The subject would have been close to the heart of a school of anthropology of the period whose members, based at UCL, often participated in rather vituperative debates about the origins and diffusion of culture. In his lectures, Dixon treated the audience to his insights on the English character, the English soul and the special characteristics of the English genius and its origins (I owe this example to Strathern 1992: 30–1).

At present, when it is still an unresolved problem to know what it is to be English, re-reading past attempts to solve the question may be more than a nostalgic indulgence. The English, Dixon declared, should be celebrated as individualists, a hybrid people, the result of 'an astonishingly mixed blend' of peoples of differing origins, producing an amalgam of talents. He devoted part of one of his lectures to providing some quantitative substance to these observations. If you could count the number of geniuses produced in each English county, then he argued, degrees of Englishness could be measured:

> The predominantly Saxon districts Middlesex, Surrey, Sussex, Berkshire and Hampshire stand low on the list of talent, comparing very unfavourably in this respect with Dorset and Somerset in the West, Buckinghamshire to the North and Kent, Norfolk and Suffolk to the East. We observe then that in the regions where the component elements are most numerous, where there is most mixed blood, the greatest ethnic complexity, genius or ability most frequently appears. The hybrids have it. Norfolk, Suffolk and Kent are the counties in which the mingling of races is greatest and precisely these counties, and not the purely Saxon or purely Norse, are richest in talent. (Dixon 1938: 108)

Reading this text now contributes to explaining why tracing cultural origins lost respectability in anthropology. Anthropologists have since by and large avoided such studies, except for a few notable exceptions in the post World War Two era

* University College London, Inaugural Lecture, Michael Rowlands, 1998.

of nation building in eastern Europe. But the consequences of the civil war in Bosnia-Hercegovina have finally put an end to even these traces. Archaeology, on the other hand, has never quite made the break with the study of origins that many of its practitioners would have wished. Whilst we are unlikely to see the word 'origins' in a book title now, nevertheless the study of prehistory is difficult to justify without some recourse to evolutionary thought. Given all these caveats, it can still be argued that origins has been a defining aporia of twentieth century social theory. Antipathy to evolutionism, diffusionism, museological views of cultures as objects measurable on some scale of complexity, and hostility to essentialist and racist claims of superior origins were necessary conditions for the development of social theory in the first quarter of the twentieth century.

In the nineteenth century, nations claimed separate origins in order to essent-ialize the distinctiveness of their right to a state. The distinction between people with or without history became institutionalized in the European academic system as a split between *Volkskunde* and *Völkerkunde*. The former were people with history, whose roots could be traced in myths, folklore and national history. The latter were those 'primitives' without history, regarded as the unchanging posses-sors of timeless cultural traits. They were the subject of ethnography and through the study of technology and material culture were objectified and frozen in the museum time of the collection and exhibition. Yet, to be included in the study of *Volkskunde* did not necessarily produce a uniform ideal of national history. Compared to English hybridity, for example, the search for Danishness after the catastrophic territorial losses of the Napoleonic wars exemplified a national quest for deep origins in prehistory. Sorenson has described one of the characteristics of being Danish as the use of the past to combine an apparent modesty/acceptance of it with a simultaneous emphasis on the unique character of the national spirit (Sorensen 1996). The loss of empire, for Danes, was compensated for by discov-ering in school education and civic action, a consciousness and pride in roots and heritage which would nurture in the young a sense of responsibility for the future (Sorenson 1996: 33). The Danish public remain today the most ardent consumers of academic books on their own prehistory in Europe. A tacit, taken for granted acceptance of the power of origins also flourishes in France. In 1985, when the French President Francois Mitterand declared at the opening of the archaeological excavations of Mont Beauvray – that *Bibracte* (the Latin name for the site) was a sacred site 'where the first act of our French history was played out', his intention was not just to claim that the nation owed its origins to a common ancestor (Vercingetorix) who resisted Roman conquest but that the resulting Gaulish/pan-Celtic identity of a hybrid population symbolized the first of many unities of racial differences in French history.

In the archaeology of knowledge, Foucault argued that the philosophical itch for origins and foundations manifested itself in conventional history in several ways

(Foucault 1984: 95). One was the linkage of the writing of continuous history to the transcendental subject; another, that there should be a symbolic logic to interpreting the past, the desire to interpret signs as representations of things. The quest for origins, he argued, was a consequence of a motivated desire to ignore the twisted fabric of history, part of a 'will to power' that writes history from an omniscient perspective where the past makes sense only when its contradictions are violently repressed. The job of interpreters of the past should not be to discover the roots of our identity but to commit oneself to their dissipation (Foucault 1984: 95). Yet, the fact is that we are confronted more than ever by the violent consequences of passionate beliefs in the essential right to claim separate religious, ethnic and cultural origins; claims which are based on the ideal of the possession of culture rather than knowing its meaning. Claims to origins in political contexts of various hues are generally a debate about possession and not about interpretation.

Moreover there are striking academic differences in response to the two principal modes of current origins discourse. Anthropologists have contributed a great deal to the rights of indigenous peoples in claims to land and other resources. The support of the rights of *indigenous* peoples against the power of the state or the economic interests of multinationals is a distinctive achievement of applied anthropology. But until recently, anthropologists have been reluctant to take seriously the claims by nationalist and ethnic separatists who base their arguments on equally essentialized ideas of culture, arguing instead for an alternative reality of motivated constructivism to explain the passions aroused.

At present the dilemma of how to adjudicate between these different claims has been contained within a not particularly satisfactory framework of cultural rights. Contemporary justifications of the cultural right to existence are based on the 1947 United Nations Convention on Human Rights which encoded respect for the rights of minorities to preserve their traditions. Liberal multiculturalism affirms a political attitude which fosters and encourages the cultural and material identities of cultural groups within a society. In part this is due to the fact that, in international law, only the State has juridical status, which has meant that human rights legislation has been organized in terms of the relationship between the individual and the state as the possessors of legal personalities. Cultural rights are personal insofar as they protect the right of the individual to a cultural life, or the right of the individual as author to protection of artistic, literary and scientific works (viz. the International Bill of Human Rights or the 1992 UN Declaration on the rights of persons belonging to national, ethnic or religious minorities)

The absence of a legal definition of cultural rights, reflects a more general problem, that of defining collective rights by contrast to those of the individual and the state. But what is striking is that all such statements assert a right to culture without defining what it is that has this right. The right to develop a culture has been asserted in several UNESCO conventions; the 1976 Algiers declaration on the Rights of Peoples refers to the right of respect for cultural identity and the 1986 Genocide Convention reaffirms the 1947 UN Convention of the right of minorities to separate cultural existence. Both the notion of social group and the idea of culture used in these conventions are quite archaic however and, from an anthropological viewpoint, ill-informed. Since the legal basis of cultural rights remains the rights of the individual as a member of a group, it is understandable that notions such as 'culture' or 'community' are used as little more than a rhetorical device. Yet the high-blown rhetoric of culture can also be used to support minority demands for separate recognition and claims to economic resources. Culture then refers primarily to collective social identities engaged in struggles for social equality and the protection of minorities from exclusion. Verena Stolcke, for example, commenting on the rise of cultural fundamentalism in Europe, has argued that this phenomenon does not just disguise older types of racism but represents a heightened concern with cultural identities that has eroded cosmopolitan hopes in the aftermath of the horrors of World War Two (Stolcke 1995).

In 1985 the European parliament convened a committee of inquiry to report on the rise of xenophobia and racism in Europe. In the same year the most well-known anthropologist of the post war period, Claude Lévi-Strauss, published the text of his public lecture delivered originally in 1971 to a UNESCO meeting on racism and discrimination. In the preface to his book *The View from Afar,* Lévi-Strauss (1985) describes the reaction of those at the meeting to the misgivings he expressed about his original statement on race and culture to the UN in 1952 in which he expressed the belief that the progress of civilization might be understood as a peaceful combination of the world's cultures into some larger pan-human combination. Changing his mind in 1971, Lévi-Strauss argued that ethnocentrism was not only in itself not a bad thing but, as long as it doesn't transform into racism, was a good thing. Whilst relative incommunicability can never justify one culture destroying another, difference in itself was not only not at all repugnant but was basic to creativity.

As usual, complete frankness was not a good idea. This reversal of attitude by Lévi-Strauss to what in a different period he had branded as essentialism, was deeply shocking to UNESCO staff. But it helped point to the contradiction which still remains unresolved in cultural rights debates – the need to reconcile the demand for creative affirmation of each identity without it leading to closure towards others. Lévi-Strauss's (1985) answer to the dilemma was to emphasize that whatever the origins of borrowing and mixture, the anthropological task is to document how each culture arrives at 'its own original synthesis' thus creating the specificity of its identity out of hybrid origins. What Richard Rorty (1980) mischievously called 'convincing one's own society that it need be responsible only to its own traditions'. Such detailed documentation, the job of the anthropologist, allowed definition with openness through the recognition of shared characteristics.

Whatever you think of this particular justification for cultural self-centredness, the agents for internationalism that Lévi-Strauss sought to shock, absorbed the lesson. For example, the year 1995, if you didn't know it, was the European Year of the Bronze Age. The European Commission spent £300,000 on promoting the idea that the origins of a culturally distinct Europe could be traced back 4,000 years to the Bronze Age when an area from the Atlantic to the Urals achieved a certain distinctive synthesis of social and technological characteristics. Equally, in the catalogue for the huge exhibition held at the Palazzo Grassi in Venice in 1991 under the title 'The Celts – the origins of Europe', the aims of the exhibition are described as 'conceived with a mind to the great impending process of the unification of Western Europe' since the Celtic Iron Age was the first historically documented civilization on a European-wide scale.

What Lévi-Strauss overlooked was that ethnocentrism without racism – assumed to be some non-racial yet natural potential for people to assert their difference – reifies common language and tradition into something more than a means to express a sense of belonging. Divesting ethnic signifiers of any particular cultural content has been a major weapon against racism in the formation of anthropology since the time of Franz Boas and others. It was precisely the fluidity of cultural identities among immigrants in the United States, supposedly demonstrating the innate plasticity of human behaviour to adapt to new circumstances, that Boas used to argue for the support of anthropologists against racist restrictions on the immigration of some ethnic groups into America. In this sense Lévi-Strauss was being adept at playing an old card in the hope it might work one more time against the essentialist argument that the possession of a unique culture was absolute and unchanging. In the American Anthropological Association statement on Human Rights in 1947, its President, Melville Herskovitz, exhorted the post-war political elite in America to respect cultural differences and to raise 'the rights of individuals to culture' as a supreme ethical value. This was to be anthropology's trump card – the subject would continue to be the advocate of indigenous minority cultures against attempts by international agencies such as the UN to universalize a set of 'Western' moral values. Herskovitz's view of holistic culture has been exposed to many criticisms over the years but no one questions his belief, inherited from Boas, that by uncoupling ethnicity, culture and identity and in particular dismantling the essentialist glue that bound them together, he had the answer to maintaining for anthropology the centrality of the study of cultural difference whilst recognizing that cultures adapt and change in response to new circumstances.

Nothing much has changed in anthropology concerning the need to uncouple race from a theory of culture, ethnicity from identity and identity from essence. There is a long tradition in anthropological thought, from Herskovitz writing on acculturation theory to Fredrik Barth writing on ethnic groups and boundaries, that has been concerned with studying ethnic difference as a signifier unrelated to any particular cultural content. There is an equally strong tradition that sees cultural categories constructed and expressed as difference mediated through social and political practice. Once uncoupled from any particular relation to ethnicity or identity, culture need no longer be viewed as a political statement about the closure of identity but rather the expression of a semiotic praxeology adaptive to circumstances and open in content.

But this resistance to essentializing arguments also entailed the risk of what might be termed the 'banality of difference'; that the proliferation of variation has the effect of rendering difference (and conflict) inconsequential. Stuart Hall has argued that by sidestepping the closures of culturally bound objects that had characterized ethnographic and folkloric research for so long, the oppositional potential of difference was co-opted, multiplied, stylized and aestheticized (Hall 1999). The proponents of multiculturalism, he feared, would co-opt the oppositional potential of ethnicity; for when difference is multiplied, relativized, exoticized and aestheticized, it is also neutralized. At a time when essentializing notions of race and identity are again on the rise and used to support efforts to close off immigration in Europe, the uncoupling of ethnicity and identity as part of the right to resist signification has a number of merits. It captures still some of the position from which Franz Boas shaped the discipline of anthropology in America during the first half of the twentieth century, a position of 'constitutive negativity' to borrow a phrase from Roland Barthes. Yet there are signs that we are reaching the end of something here. For a number of writers, new forms of ethnicized violence have been associated with a profound change in the forging of localized identities in the late twentieth century. Fears of globalizing homogeneity are matched by passions over the preservation and elaboration of localized identities. Attacks on immigrants in the new fortress Europe are embedded in identity politics which stress sameness and repetition. Olsen (2001) has described the consequences of this in Scandinavia as the ludicrous paradox that Norwegians will stress their continuity and essential sameness with Viking ancestors who are more 'foreign' to their modern descendants than any modern Tamil, Pakistani or Vietnamese immigrant 'stranger' living in their midst. Sameness and continuity – the nineteenth century paradigm of continuity that early anthropology in part reorganized its concept of culture to combat – is vitalized both as a framework for identity and roots and in some cases as a means to resist processes of fragmentation and disruption in the present. This is obviously related to contemporary structures of power and hegemony as well as representing the limited resources available to minorities to resist centuries of dispossession and marginality. The fact is that a similar essentialist rhetoric can be used by indigenous groups both to resist domination and as a means to reinforce indigenism as a reactionary museum image forced on ethnic minorities by academics and politicians. In the prison house of tradition and authenticity any change may be seen as a contamination and loss of identity.

Perhaps another sign of the end of something is the identification of cultural rights with the possession of cultural property – a new kind of materialization of identity (or perhaps the recreation of a very old one). There is after all a long held belief that objects can be said to stand for cultures and the amount you have of 'your own' cultural objects (cultural heritage) is a measure in some sense of relative cultural complexity. In a populist and often violent response to a fear of cultural loss, ethnic violence in recent years has been as much directed towards the destruction of cultures as collections of objects as against the bodies of their subjects.

Culture and Property

On 6 December 1992 at Ayodhya in North India, watched by politicians who are now in the coalition cabinet led by the BJP, a mosque was destroyed by militant Hindus who climbed up the three domes of the sixteenth century building to demolish the building. Ten years later, the ruins remain guarded and the plans to build a temple on the site of the mosque remain a commitment, if deferred, of the government. Ayodhya, a town revered by Hindus as the birthplace of Lord Rama, became a focal point for the rise to government of the right-wing Hindu nationalist Bharatiya Janata Party. The mosque it is alleged by the Hindu World Congress was built in 1528 on the orders of Babar, the first Muslim Mogul of Emperor of India, after he ordered the demolition of a temple on the same site dedicated to the birthplace of Lord Rama. Parts of the original temple, it is alleged, were incorporated into the building of the mosque and some Indian archaeologists have carried out excavations which they claim as evidence for the prior existence of a temple on the site (cf. Layton, Thomas and Stone 2001).

The struggle to claim the site for either a mosque or a temple is fed by feelings of redemption and retribution. If Muslim rulers inflicted their tyranny on Hindus then Hindus have the right to inflict it on them today. If Hindu temples have been destroyed in the past then mosques will be destroyed today. The demand is for re-enactment. If conquerors asserted their power in the past by destroying religious buildings, then liberation now depends on destroying their buildings in the present. The construction of such a Hindu consciousness and identity, served by the unification tradition, is the goal of the BJP even though Ayodhya has been a sacred site for many religions and is a place where various forms of religious worship had flourished concurrently. It had been an important religious centre for all three religions of early India, Buddhism, Jainism and Hinduisim. From about the fifth century BC a fairly large Buddhist community was present there. According to Jain tradition, Ayodhya was visited by the founders of Jainism and became a centre for Jain pilgrimage.

Ayodhya is therefore certainly a contested site, and is the subject of a mythic narrative which details the destruction and restoration of a shrine. Temples figure prominently in Hindu Nationalist rhetoric as part of a broader historical vision of India facing barbaric foreign invasions, first Muslims and later the British. The BJP's official White Paper on the future of cultural heritage in India supports the case for building a new temple on the site of the demolished mosque at Ayodhya – by saying

> here was an ancient temple which had been ravaged, looted and razed to the ground repeatedly by foreign invaders. Every time the temple was razed to the ground and a mosque put in its place by the marauders, it sprouted again – only to be pulled down again. (BJPO 1993: 16)

The BJP's use of the organic metaphor, the ever-sprouting temple, renders it and its restoration to be part of nature – the irresistible natural expression of the 'urge' of the Hindu 'race'. The idea of collective memory, lodged in grievance, gives iconography a fixed and unchanging quality. But collective memory has to be given form – narrated and re-narrated – and it has to be consistent with its time and period. Hence the building and rebuilding of temples, rather than their preservation, is the material objectification of this type of remembering. The legacy of a British colonial sensibility of conservation and a radical Hindu notion of destruction and rebuilding can become entangled in the same site. Ayodhya at the time of its destruction was protected by antiquities legislation first promulgated by the British. As part of their interest in controlling and conserving the Indian past, the British had instituted an archaeological survey of India and one of the great promoters of this Survey, George Nathaniel Curzon, British Viceroy from 1899 to 1905, described the instituting of the Survey and spoke of its purpose as a comprehensive project of knowledge. One of the tasks he established, as part of the archive, was to cherish and conserve sites of historic and artistic interest in India as museums. Ancient temples that had fallen into disuse would find new life as museums; protected sites for tourists and students of India's past. What has been assembled as 'evidence' for the origins of Ayodhya as a Hindu temple is therefore the ideological product of a view of the past that can imbue the classificatory knowledge of the archive with the emotional certitude of Hindu cosmology. The clash and collusion of archaeologists and nationalists in this particular case does represent something more basic in late twentieth century attitudes towards identity. The former intends to preserve the past, the other wishes to re-enact it. One wants to preserve a notion of cultural rights embedded in cultural property – the other reworks an essentialist notion of cultural uniqueness. If separated, one can be used to evaluate or enthuse the other; combined they can justify the committing of horrific acts.

The concept of culture being used in these contestatory moments is much less straightforward than it used to be. Culture as evidence of diversity in human thought and ideas is no longer sufficient for coping with these kinds of antagonism. It assumes too much about uniqueness and considers diversity as measurable on a similarity in scale. A concept of culture is assumed to mean something that can be represented in common patterns and configurations of beliefs and materialized as specific relations between elements, symbols, behaviours, practices or structures of meaning.

This contrasts with an older, rather nineteenth century museological view of culture, which treated it as an object rather than a practice. In this view, the meaning of culture was contained in its material form rather than in its practice; the meaning was in the possession of culture as a 'thing' rather than in the process of its production. Michael Lambek has disparagingly called this a contents and container view of culture, but for the Hindu nationalists, the particular form of identification they advocate is grounded in a similar materialist notion of the temple as a thing, on the animist principle that life is constantly breathed into the inert through the actions of the devoted. A similar argument lies behind current demands for cultural repatriation of bones, artefacts and sometimes whole collections of colonially plundered objects from western museums and academic institutions. When the trustees of the Kelvingrove Museum in Glasgow agreed to return the ghost shirt of a dead Sioux Indian to a nation claiming its cultural property – perhaps there was a relief in avoiding a problem but equally there was a will to recognize the argument that an object may have a life of its own; that, whatever the legal rights involved, the ethically correct action would be to return the object to where it belongs. The ethic is revitalist, in the sense that loss and grievance need to be remedied and restored through material acts such as the return of lost or stolen cultural property. On this basis, the possession of a culture is redemptist; you can have either more or less 'culture' on which to build respect in the eyes of others. To accomplish this requires in-depth study of the local and the particular, the creation of museums and archives to sustain identity and the use of ethnographic and archaeological methods to excavate the sedimented cultural materials on which secure identity resides.

But if identity has been accumulated in this way, then it is equally the case that it has been the object of destruction and restoration. In Sarajevo 1995, after the bombardment of Dubrovnik, the leader of the Bosnian Serb forces asked 'What's all the fuss about some old stones. We shall build Dubrovnik even more beautiful and older than before.' In Bosnia-Hercegovina between 1991–1995, over 85 per cent of the Muslim architecture was either destroyed or severely damaged beyond the point of repair. In the mid nineteenth century, Belgrade was the largest Turkish city in Europe with over 52 mosques and 22 Koranic schools. There are no minarets and few mosques left now. And the destruction seems not to have been vented on muslims as aliens. As a Bosnian Serb schoolteacher in ethnically cleansed Zvornik in Eastern Bosnia was reported to have said to Marian Wenzel of the Bosnia-Hercegovina Heritage Rescue Group:

> Who are the Muslims? I will tell you. They are Serbs converted to Islam. So we don't have trouble with Turks. We have trouble with bad Serbs. (Wenzel personal communication)

Making bad Serbs good was and still is a bloody business. But cleansing has its contradictions. Whilst there may not be many mosques left, Muslim gravestones and cemeteries do survive. Good and bad Serbs, like good and bad Croats, have deep respect for the dead and this is probably an active feature of funerary rites in Bosnia dating from pre-Christian Vlach society. The demeaning of Muslim culture in Bosnia – widely regarded in the nineteenth century as the best of Europeanized Ottoman Culture – made it easy for the Bosnian Serbs to almost completely wipe it out in one blow. Already this had begun during the Austro-Hungarian period when the Austrians rewrote Bosnian history to make people feel either more Croat or Serbian (i.e. Catholic or Orthodox) rather than Bosnian (and Muslim). The work of ripping down and throwing away evidence of Ottoman culture that started then has for many now simply reached a final completion. As told recently to Marian Wenzel by a member of the Serbian Academy of Sciences in Belgrade.

> All this is only noticed internationally because we didn't get around to eliminating these mosques before. We were involved in other things. For instance there were Muslims in Hungary almost the same length of time as here. Do you see any mosques there? Of course not. They were smarter than we were and got on with it at once. Evidence of their Turkish invaders was removed quickly and quietly a long time ago. (Wenzel personal communication)

In the chaos that has ensued since the Dayton Accord, no respect has necessarily been shown for what has been left. Current restoration of mosques, heavily financed by Saudi Arabia, is being carried out on condition that Ottoman arabesque wall painting, thought decadent by Saudi sponsored Wahabism, be painted over. What art historians had previously recognized as Bosnian styles in gold metalwork of the fifteenth century and in stonework originating as a Gothic internationalizing style of the eleventh century, have recently been redesignated as Bosnian Croatian at a convocation on Bosnian art. The objectification of culture, redesignated and filled with differing elements, is hard at work in war torn Bosnia, classifying and identifying people in terms of homogenous origins; reworking the project of ethnic mapping for a future resolution of historical injustice.

Diversity is generally presented in cultural terms with the relevant markers for culture described in terms of bounded identities. In the case of Bosnia, for example, one consequence of the Dayton Accord has been the acceptance of Serb and Croat as the key cultural designators of identity in the area at the expense of hybrid categories and in particular the predominance of Bosnian Muslims in the population. The demand to essentialize as 'marked groups', Croat and Serbian ethnicities in post-Dayton Bosnia is justified by the increasingly undesignated identification of being Muslim in Bosnia.

Phrases, such as 'we are a nation because we have a culture', or 'this nation has a long history' (and by implication you as outsider cannot understand it) are metaphors of possession that work only to the extent that it is known what it means to be dispossessed. In material culture terms this inversion of what was a dominant Muslim culture, in terms of cultural heritage and the way such things are inscribed on memory, occurred not only through acts of physical destruction of objects and persons but also by reconstituting the identification of named people to land and language. Current tendencies to combine 'culture' and 'ethnicity', promoting cultures as real things to live and die for, has encouraged a new scholarly consensus against essentialist approaches to culture and obliged the academy to work against such tendencies wherever detected. But, as we have seen, ethnic labels are now being validated and defined as homogeneous and discrete cultures of shared cultural heritage. The Boasian/Herskovitzian insistence that race, language and culture vary independently of each other – the assumptions that formed the post-war consensus in anthropology on how to conceptualize culture – is deeply compromised by current ethnic/cultural politics that materialize the past in terms of heritage and cultural property.

From this perspective, cultures are a product of heritage because like objects they can be identified and appropriated according to their physical traits. Cultural security and identity is assured by the act of possession. And substantive heritage is normative, predictive of individuals, behaviour and ultimately seen as the cause of why those who have it, behave as they do. But turning cultures into objects can only be legitimized through the conversion of possession of culture into a right and these are rights in cultural property

Cultural Rights and Property

The general and rather depressing point of my discussion so far concerns the destructive power of cultural origins. Cultural property as heritage refers to something more than a Western ethnocentric notion of property as a relation between persons through the exclusion of things. It suggests a more general point that physical killing is one aspect of the material eradication of cultural integrity, and the latter is part of the process by which people are dehumanized and transformed into suitable objects for destruction. If the rise of intense and localized forms of ethnic violence has characterized the late twentieth century then it may not be insignificant that it has also been accompanied by the rise of an intense concern for cultural heritage and a contested politics of cultural rights.

Defining cultural property as a right means seeing heritage as part of the process of reproducing social relationships rather than simply the accumulation of things. Yet western notions of cultural property are embedded in particular kinds of individual relationship, e.g. the right to exclude others from the possession of things. As a relationship between persons and things it is associated historically with the rise of capitalism in the West and its dependence on private property as individualized possession.

When nationalists describe French Canada as a nation with a distinct identity, a pre-emptive claim is made to knowledge because they are 'haunted by a vision of totality' that can only be achieved when a people become 'an irreducible, homogenous unit, securely in control of its borders, self-contained, autonomous and complete' (Handler 1988: 194). And nationalists are only too well aware that academics are not necessarily their best supporters in this. The Hawaiian nationalist Haunani-Kay Trask (1991: 162) recently characterized anthropologists and historians as 'part of a colonising horde because they seek to take away from us the power to define who and what we are, and how we should behave politically and culturally'. Ethnic nations and groups prove their existence and gain respect by conserving and preserving their property. Hence the proliferation of museums and conservation legislation and competition between them to demonstrate that they have truly world class heritages.

A relationship is established therefore between completeness and possession. How this quest for authenticity encourages, static, unchanging images of identity has been much commented upon. As Richard Handler has demonstrated, the preservation of patrimony in Quebec, whereby a body of objects, buildings and sites became defined, inventoried, classified and enclosed as authentic 'Quebecois culture', was not the result of academic studies and reports but of city administration, political policy, legislation and heritage management. Heritage is constructed as an archive through processes of naming and classifying, documenting and recording the details of what is to be restored or conserved. It is a product of the more general bureaucratic functions of the state which gives it the appearance of an identity frozen in time. Yet contrary to the dry, rational procedures of heritage management, it is precisely the ethnic activists' belief that their group's survival depends upon the secure possession of a culture that renders debates about who owns the past so passionate. The terms culture and history become more or less synonymous. This is because a group's identity is said to be preserved and embodied in material artefact such as buildings, works of art and antiquities and confirmed by the groups undisputed possession of them as its cultural property. What would be termed by many as an old fashioned museological view of culture – associated in anthropology and archaeology with outmoded culture historical approaches – is revealed as a basic tenet of international legislation on cultural rights. The right of the individual to live within a particular way of life underlies contemporary multicultural rights legislation – e.g. as in the application of the concept of fairness in Rawls' theory of justice. Equally the 1976 declaration by the UNESCO panel on restitution that 'cultural property is a basic element of a people's identity' and 'being depends upon having' is a pretty good statement of the relevance of modern possessive individualism to the international politics of culture.

A discourse perspective would, I imagine, quickly pick up the underlying codes shared by these three apparently contrary instances of identity politics. In a parallel influential study of organizational factors in overseas aid projects, James Ferguson has argued that:

> the thoughts and actions of development bureaucrats are powerfully shaped by the world of acceptable statements and utterances within which they live; and what they do and do not do is a product not only of the interests of various nations, classes or international agencies but also, and at the same time, of a working out of this complex structure of knowledge. (Ferguson 1994)

He demonstrates how only certain, in particular technicist statements, enter into the discourse of development agencies and the World Bank. Analogues can be drawn with UNESCO discourse on cultural rights and the protection of cultural property which can be equally determinist in defining cultural heritage as an object of technical support in development and thereby excluding from consideration some of the fundamental political factors, such as national ideologies, that are its principal support. What has been described as a museological view of culture is spreading rapidly through various forms of legislation supporting cultural rights and restitution of cultural property legislation and the organizations set up to administer them (cf. Niec 1998). By equating loss of possession with a deprivation of being, cultural heritage is assumed to be the collective possession of a country or group of origin regardless of any mode of transactions or forms of legality. A quite explicit part of this argument is the belief that the cultural right to heritage is fundamentally a right to the ownership of all forms of culture as intellectual property. This can and has resulted in a huge matrix of all kinds of things in which people can exercise rights (land, designs, language, magical spells, myths, songs, etc) as a form of copyright (Brown 1998). To some the very idea of property would be inappropriate as a way of describing cultural rights. It is simply the imposition of a Western category, through international law, which distorts indigenous realities. From the perspective of international law, collectivities should be treated in the same manner as individuals. Cultures are identified with property and an unproblematic equation is made between loss of property and diminishment of the person and their culture. Here the opposition between private, personal interests and universal claims to the common good, is understood to cover all eventualities.

To counter this tendency to homogenise identity, there is at present a burgeoning anthropological literature concerned to expose the strangeness of these Western notions of property. Strathern for example contrasts Western concepts of property to the New Guinea Highlands, where

> equations between women and wealth – have little to do with property in the sense of rights over objects. They are related instead to the way in which wealth items signify aspects of the person. (Strathern 1988: 229)

The use of objects, she argues, often signifies relations of the person to others in such a way that the social relation is made evident at the same time as the right. In the light of many such examples of the ways in which people exercise rights in relation to each other that would not necessarily constitute property, it might well be asked whether there is any point in continuing with the debate. In this view the very idea of property is inappropriate and serves only to distort indigenous realities.

One answer is that probably there is no choice in the matter. The link between cultural rights and property is deeply embedded in current and proposed international legislation. For example, the UNESCO Draft Declaration on the Rights of Indigenous Peoples contains the following resolutions presented to them at various National and International meetings:

1. indigenous peoples should be recognized as the primary guardians and interpreters of their cultures, arts and sciences whether created in the past or developed by them in the future.
2. all forms of tourism based on indigenous peoples heritage to be restricted to activities which have the approval of the peoples and communities concerned and which are conducted under their supervision and control.
3. protection against debasement of culturally significant items.
4. appropriate standards to be presented for the use of indigenous imagery by the tourist industry.

In other words, it is little more than a list of culture property claims which should be protected. Michael Brown (1998) also cites the case of the Inter-Apache policy on Repatriation – a declaration issued by a consortium of Apache tribes demanding exclusive decisio-making power and control over Apache cultural property, which they define as:

Cultural property includes all cultural items and all images, text, ceremonies, music, songs, stories, symbols, beliefs, customs, ideas and other physical and spiritual objects and concepts inalienably linked to the history and culture of one or more Apache tribes. This term includes all intellectual and ally representations and reproductions, by any means, of any Apache cultural property. [i.e. encompassing ethnographic field notes, feature films, historical works, etc].

Brown implies that such encyclopaedic demands are usually opening gambits in the negotiation between Native American peoples and Anglo-American institutions. Yet before negotiation can take place certain assumptions must be in play:

1. An ethnic nation – a people – is agreed to have enduring, comprehensive rights in its own cultural production and ideas including the right to control their representation in whatever media.
2. A group's moral right to its own ideas and practices is a form of ownership, the equivalent to European ideas of authorship and copyrighting.
3. These rights are not analogous to copyright in intellectual property, where rights of authorship are held for a limited period precisely in order to encourage the free circulation of ideas by providing adequate remuneration to authors. Instead cultural ownership is conceived in perpetuity and cultural property should be transmitted (and not transacted) only as an aspect of belonging to an indigenous culture.

This of course allows little room for the establishment of mutual respect in shared cultural property. An emphasis on the unique character of cultural property, as something peculiar to the identity of one homogeneous cultural grouping, weighs against the principles of fairness basic to current multiculturalist legislation. The implication of a nineteenth century museum based, objectified and static version of culture being reintroduced under the guise of the saving the cultural property and identity of a marginalized and threatened indigenous community has real possibilities of bringing us back one more into the prison house of tradition.

Yet few are going to dispute the right of indigenous peoples to protect their cultural heritage and prevent it being exploited. We can also understand why museum professionals and others would prefer the development of frameworks based on joint stewardship rather than legal models based on rights and ownership. Museums and archives holding collections of indigenous material are more likely now to enter into negotiation over shared access to inalienable property, making joint arrangements of mutual benefit than argue fruitlessly over who ultimately owns cultural property. But the other side of that coin is why should not the cultural knowledge of cultural minorities be treated by its owners as a commodity? Outside the walls of the museum, tourism and heritage management are more than ever entwined in the process of the right to use one's cultural heritage as a resource in sustainable development. Tourism and tourist art or the naturalizing of cultural/ natural heritage as an experience does commodify culture but there may be nothing particularly wrong with this. Does this offend the spirit of repatriating inalienable cultural property to their original owners, held not by an indigenous people but by the Western-trained museum curator or conservator who basically see their role as preventing the commodification of culture?

Anthropologists are already heavily involved in ameliorating the social impact of development on indigenous peoples and cultures. In the most well-known case of Australian Aboriginal land-right claims, the destruction of sacred sites went hand in hand with the gradual dispossession of Aborigines from their traditional lands. Asking for recompense in order to permit further damage to take place through mining or allowing tourist access to restricted territory has caused great controversy along the lines of how you attach a price to allowing the destruction of a sacred site. Abstract answers do not really help – and negotiating case by case over compensation takes up two issues which are of much wider significance for compensation to indigenous peoples. First, the latter are only in this position of negotiation because of hard-fought battles to claim ownership of land and sacred sites as a right contested in court. Secondly, negotiation takes place when some issue of public responsibility and accountability is at stake. And in order for the debate not to degenerate into accusations of using claims to cultural property as a convenient lever for monetary gain, the integrity of sites, objects and archives protected must be beyond doubt and, furthermore, this has to be recognized by all parties concerned. What might be seen as an emerging public policy anthropology over issues of compensation, repatriation and restitution of cultural property offers no hard and fast prescriptions for making these decisions at present. It obviously provides major opportunities for anthropology and its recognition in the public domain although not without attracting criticisms as to the impartiality and critical abilities of its practitioners. The issue remains how to collaborate and how to negotiate and how to participate with greater political realism.

Conclusion

At present we are witnessing the growth of an identity politics that encourages the revival of older views of culture which for many anthropologists were thought to be truly dead and buried. Origins is not a fashionable buzz-word at present but despite many wishing it was not the case, has still remained a central idea around which cultural identity continues to be mobilized.

I have linked this to the rapid proliferation of a global discourse of cultural rights and property that encourages the political identification of local movements with a language of identity and roots. All three examples discussed in this chapter – ethnic nationalism, multiculturalism and indigenous movements – have adopted culture history as a means of resisting processes of fragmentation in the present and of giving people back their identity. What many would see as a Eurocentric and nationalist view of the past, is now widely influential, even among people for long at odds with westernisation. It is not difficult to see why indigenous peoples or ethnic minorities might wish to define themselves in terms of common cultural experience and the possession of a common past in their struggle against exploitation, marginalization and dispossession nor why anthropologists have been often heavily involved in underpinning the contemporary political and legal recognition of these interests.

In many respects what is termed 'strategic essentialism' as a way of defining cultural rights that would be acceptable in international political terms, has been denied significance by anthropological views of culture that were actively hostile to tracing forms of present–past continuity. The propensity to see ethnic boundaries as socially constructed in the present has also alienated some aspects of anthropological practice from the demands of indigenous and ethnic minorities. Anthropologists, for so long inured against any hint of a culture-historicizing framework, have enormous difficulties it seems in understanding the passions generated in the often emotional support for ethnically phrased political projects. Attempts to account for ethnicity solely through situational or presentist types of analysis fail to give adequate consideration to the taken for granted commitment to identities by people who are passionate about their pasts.

Anthropologists are presented with many problems in coming to terms with these 'elements of heritage' on the move. As seen, international policy forums dealing with the protection of cultural knowledge, tend to apply the language of property as a solution to all problems of ownership. Cultural heritage is a term that links identity with the past in a quite unproblematic manner in much international legislation and yet in any particular context requires systematic analysis and disentangling in order to show how it is being used to define relations between persons. Marilyn Strathern reminds us that there is an inevitable link, at least in the modern context, between issues of identity and matters of property (Strathern 1999: 134). Yet the term 'property', whilst making reference to relationships, does so through the mediation of things. For anthropologists concerned with cultural difference, aversion to such universalizing tendencies is deeply engrained and yet, as seen, it is no longer possible to take an either/or position on such questions. Instead, it is a matter of analysing which aspects of culture are property like and which are not in situations which involve, to use Hann's (1998: 7) term, a 'distribution of social entitlements'.

The idea that cultural heritage represents a system of relationships has the advantage, as Strathern reminds us, of avoiding the binary categories of public/private, collective/individual that notions of property lead us into. It encourages the belief that indigenous people and professional museum curators may be 'guardians' of heritage and share responsibilities in a moral activity rather than disputes over ownership. This is clearly the way things are going at present, but it is by no means certain that this will not be swamped by the sheer scale and complexity of globalization. The massive expansion of international legislation and the institutions required to administer the protection of cultural and intellectual property imply global solutions that will have to be made relevant for each particular case. It seems most unlikely that this will happen and instead legal regimes will be deployed to mediate with and represent local interests. In such a contested world of international lawyers, policy experts and consultants, it seems doubtful that the particular merits of cases based on moral custody are likely to have much of a voice. But the extent to which anthropologists are involved and whose interests they represent, remains an open question which should concern us all.

References

BJPO (1993), *White Paper on Ayodhya and the Rama temple movement*, New Delhi.

Brown, M. (1998), 'Can Culture be Copyrighted?', *Cultural Anthropology* 39: 193–222.

Dixon, W.M. (1938), *The Englishman*, London: Hodder and Stoughton.

Ferguson, J. (1994), *The Anti-Politics Machine* Minneapolis: University of Minnnesota Press.

Foucault, M. (1984), *The Archaeology of Knowledge*, London: Tavistock.

Hall, S. (1999), 'Whose Heritage', *Third Text* 49: 3–13.

Hann, C. (1998), *Property Relations: renewing the anthropological tradition*, Cambridge: Cambridge University Press.

Handler, R. (1988), *Nationalism and the Politics of Culture in Quebec*, Madison: University of Wisconsin Press.

Layton, R., Stone P.G. and Thomas, J. (2001), *Destruction and Conservation of Cultural property*, London: Routledge.

Lévi-Strauss, C. (1985), *The View from Afar*, Oxford: Blackwell.

Niec, H. (1998), *Cultural Rights and Wrongs*, Paris: UNESCO.

Olsen, B. (2001), 'The End of History? Archaeology and the politics of identity in a globalised world', in Layton R., Stone P.G. and Thomas J. *op cit.*

Rorty, R. (1980), *Philosophy and the Mirror of nature*, Oxford: Blackwell.

Sorenson M-L.(1996), 'Nationalism and Danish Archaeology', in M. Diaz-Andreu and T. Champion (eds), *Nationalism and Archaeology in Europe*, London: UCL Press.

Strathern, M. (1988), *The Gender of the Gift*, Cambridge: Cambridge University Press.

Strathern, M. (1992), *After Nature*, Cambridge: Cambridge University Press.

Strathern, M. (1999), *Property, Substance and Effect*, London: Athlone Press.

Stolcke, V. (1995), 'Talking Culture: new boundaries, new rhetorics of exclusion in Europe', *Current Anthropology* 36: 1–24.

Trask, H.-K. (1987), 'Response to Keesing', *Current Anthropology*, 28, 2: 171.

—— (1991), 'Natives and Anthropologists: the colonial struggle', *The Contemporary Pacific* Spring: 159–76.

–6–

Landscape and Politics
Barbara Bender

Introduction: A 'Western' Point of View

Ask anyone brought up in, or influenced by, Western ways of thought what they think landscape is about and they are likely to answer in terms of 'views'. Something seen, usually at some distance. Often beautiful, usually rural, or if not – as in 'an industrial landscape' – then with a value judgement attached (*not* beautiful/*not* rural).

Let us start by considering this Western 'viewpoint'. It was in Europe that the word 'landscape' originated. First, there was the medieval German word *Landschaft* which referred to a feudal peasant land-holding – a small, familiar place. This usage disappeared and when the word 'landscape' re-emerged, in Italy and the Low Countries, in the seventeenth and eighteenth centuries, it meant something quite different, something intimately connected with people's changing understandings and negotiations of a mercantile Renaissance world (Cosgrove 1984). The way landscape came to be understood was part of a wider process of individuating people, of separating out people–culture and land–nature, and of asserting control over land–nature. There was also a gendering: people–culture, male; land–nature, female. Landscapes were to be 'looked at' or 'looked over'; the value of land as commodity went hand in hand with the value of land as an aesthetic, beauty was in the eye of the be-*holder*. From an 'elevated viewpoint' the gaze swept the landscape, first in paintings, then in ways of positioning oneself within the landscape, and finally, through material intercessions to make the landscape conform to the aesthetic (Bender 1993).

These are, of course, class privileged ways of seeing the world. We need to ask not only about the interaction between particular social, economic and political conditions and ways of engaging with the land but more precisely about *who* is doing the interacting, and how. Or rather, who *appears* to be doing the interacting – where, in this scheme of things, are the labourers whose villages are removed to make way for the landscaped estate and whose physical labour is airbrushed from park and garden?

We might also want to question further and to ascertain that these rural landscapes are interdependent upon other landscapes – of city, factory and plantation (Williams 1973). Jane Austen, for example, in *Mansfield Park*, explored the nuanced landscapes of an absentee landlord, a nouveau riche couple, a modest country gentleman and a woman of limited means (Daniels and Cosgrove 1993), but she failed to mention the overseas slave plantations that underwrote the fortunes of the 'good' absentee landlord (Said 1989).

So one way to understand the Western concept of landscape is in terms of historically and geographically specific socio-economic, political and cultural relations. While much has changed in the western world since the word 'landscape' was coined, we retain in large measure the nature/culture divide and the assumption that nature is passive and we are in control. We still look out over . . .

But we also need to integrate into the discussion of socio-political relations the processes of socialization by which people come to understand the world around them: people make landscapes, but also landscapes make them. From childhood onwards people negotiate space and place, learning through being told, through emulation and almost subconscious habit, what is permissible and what is not. Learning where to go, and when, and what is appropriate to different places at different times and in different contexts (Giddens 1985, Pred 1990). This is a negotiated process, one open to question and with the potential for change.

In thinking about these processes, it becomes apparent that landscapes are not just 'views' but intimate encounters. They are not just about seeing, but about *experiencing* with all the senses. An experiental or phenomenological approach allows us to consider how we move around, how we attach meaning to places, entwining them with memories, histories and stories, creating a sense of belonging (Tilley 1994, Feld and Basso 1996, Basso 1996, Sutton 1998).

But not all places are familiar and rooted – not now, and not in earlier time periods. People move or are moved (Gosh 1992). We have to encompass the experiences of travellers, traders, refugees, migrant workers . . . Landscapes can be unfamiliar, fearful, even hateful and alienating (Bender and Winer 2001). Augé (1995) coined the phrase 'non-places' to denote the arid travelling spaces of globalization, but, in reality, there are no non-places. However arid or vapid they are always some – place.

Landscapes can never stay still – feelings and engagement with place and landscape are always in the making. Nor can they be situated only in the present, for they contain and are referenced on what has gone before. People sometimes talk about the landscape as 'palimpsest', meaning by this that past activities leave their signatures upon the land. But this suggests that elements of the past are simply left 'in place'. In reality, things are more complicated: people invest these elements with new meaning, they re-use them literally or figuratively. Or they neglect or forget them. What is left out of the story is often as interesting as what left in (Bender 1999).

Encountering the Western View-point; Other Ways of Being in the World

We have seen that landscapes are experimental and porous, nested and open-ended. They mean different things to different people, and some people's sense of place and landscape are validated, some marginalized. We have seen that 'Western' landscapes are culturally and historically particular, that they are intimately linked to the development and expansion of capital and that they open out towards, and are often dependent upon, other landscapes in other parts of the world. We could follow the fortunes of the European 'view-point' as it negotiates colonial expansion (Mitchell 1989, Pratt 1985, Dubow 2001), or, more recently, other forms of global encounter – including tourism (Selwyn 1996). In so doing we watch the European way of 'being in the world' come into conflict with other, quite different understandings of landscape.

Which, belatedly, brings us to other ways of being in the world. Although the word 'landscape' originated within a European context it needs to be released from this historical and geographical straitjacket. Everyone, everywhere, perceives, negotiates, interacts with the material world around them, and we need to try to understand the multiplicity and complexity of people's engagement with place and 'scape (Hirsch and O'Hanlon 1994; Bender 1993). Anywhere, everywhere, people understand their world in subtle, contradictory and changeable ways. Anywhere/everywhere some people's understandings are valued more highly than others. Anthropologists have often focused on 'traditional' ways of being in the world, emphasizing the way in which people and landscape are intimately beholden to each other; the way in which the western nature/culture divide is inoperative and the relationship is active in both directions. The emphasis has been on the creation of familiar places, on the inflection of place in memory, myth and story. But even in small-scale societies people move around and are in contact with less familiar places or unfamiliar faces and often people's sense of place is charged, or made more self-conscious, through contact and confrontation (Morphy 1993, Strang 2001, Rowlands in this volume). Most particularly in the uneasy world of capital penetration versus indigenous land claim, taken-for-granteds on both sides have to be spelt out. For the indigenous people, there is not simply an incompatibility of 'view-points' but also the need to reconnoitre and circumvent judicial and administrative systems geared to very different and alien relationships to the land.

The Anthropology Department at University College London

In the Anthropology Department at University College London, lecturers and post-graduate students have worked on landscape – recognizing it as something open-ended, polysemic, untidy, contestational and almost infinitely variable. Some have worked in the context of small-scale societies (Küchler 1993, Morphy 1993); or of prehistoric societies (Tilley 1999, Bender, Hamilton and Tilley 1997). Some have moved towards a phenomenological approach, recognizing that landscape is not just about seeing but about a bodily immersion involving all the senses (Tilley 1994). Some have considered the materiality of memory – the way in which places and paths have stories, myths, legends hung around them. Or the way in which forgetting and erasure may be as important as remembering (Küchler 1993). There have been questions about time and landscape (Bender in press); about landscape and a sense of identity (La Violette forthcoming, Broughton forthcoming). Discussions of contestation (Bender 1998, Garner 2001), and about heritage landscapes, and the way in which landscapes are presented and exhibited (Basu forthcoming, Tilley 1999). Recently, some of the debate has moved away from familiar landscapes to landscapes of movement, exile and diaspora (Bender and Winer 2001, Basu 2001).

The chapter that follows is part of a book, *Stonehenge: Making Space*. The book arose out of a contemporary confrontation at Stonehenge between Government and Heritage managers on the one side and New Age travellers and free festivalers on the other. It seemed important to try and understand how people have engaged with this landscape, not just now but through time, and how, from the beginning, it was contested. How the contestations are never equal but forged through the inequalities of power. The book spanned over four thousand years of changing involvement by those who built and later by those who used, ignored or even destroyed the stones and the surrounding landscape. The chapter reproduced here ends with the current confrontation. In the process of getting to know some of the groups I became involved in an attempt to open the debate towards a wider audience. Working with free festivalers, Druids, local people, tourists and so on, we put together a travelling exhibition which questioned some of the conventional wisdom about place and past, and created a space in which many different voices could be heard and where people could reaction and question.

I would like to end by suggesting that, where possible and appropriate, anthropological insights can and should move into the public arena. There was a time when academics sniffed at 'applied' anthropology. But that time is over, and we have the right and the duty to offer, if not answers then, at least, insights.

References

Augé, M. (1995), *Non-Places: Introduction to an Anthropology of Supermodernity*, London: Verso.

Basso, K. (1996), *Wisdom Sits in Places*, Albuquerque: University of New Mexico Press.

Basu, P. (2001), 'Hunting down home: reflections on Homeland and the search for identity in the Scottish Diaspora', in B. Bender and M. Winer (eds), *Contested Landscapes: Movement, Exile and Place*, Oxford: Berg, 333–48.

—— (forthcoming), in C. Tilley (ed.), *Material Culture and Social Identity*.

Bender, B. (1993), *Landscape: Politics and Perspectives*, Oxford: Berg.

—— (1998), *Stonehenge: Making Space*, Oxford: Berg.

—— (1999), 'Subverting the Western Gaze: mapping alternative worlds', in P. Ucko and R. Layton (eds), *The Archaeology and Anthropology of Landscape*, London: Routledge.

—— (2001), 'Landscapes on-the-move', *Journal of Social Archaeology* 1(1) 75–89.

—— (in press), 'Time and Landscape', *Current Anthropology.*

Bender, B., Hamilton, S. and Tilley, C. (1997), 'Leskernick: stone worlds; alternative narratives; nested landscapes', *Proceedings of the Prehistoric Society* 63: 147–78.

Bender, B. and Winer, M. (eds) (2001), *Contested Landscapes: Movement, Exile and Place*, Oxford: Berg.

Broughton, H. (forthcoming) in C. Tilley (ed.), *Material Culture and Social Identity*.

Cosgrove, D. (1984), *Social Formation and Symbolic Landscape*, London: Croom Helm.

Daniels S. and Cosgrove, C. (1993), 'Spectacle and text: landscape metaphors in cultural geography', in J. Duncan and D. Ley (eds) *Place/Culture/Representation*, London: Routledge.

Dubow, J. (2001), 'Rites of Passage: travel and the materiality of vision at the Cape of Good Hope', in B. Bender and M. Winer (eds), *Contested Landscapes: Movement, Exile and Place*, Oxford: Berg.

Feld, S. and Basso, K. (1996), *Senses of Place*, Santa Fe: School of American Research Press.

Garner, A. (2001), 'Whose New Forest? Making place on the urban/rural fringe', in B. Bender and M. Winer (eds), *Contested Landscapes: Movement, Exile and Place*, Oxford: Berg.

—— (forthcoming) in C. Tilley (ed.), *Material Culture and Social Identity*.

Giddens, A. (1985), 'Time, space and regionalisation', in D. Gregory and J. Urry (eds), *Social Relations and Spatial Structures*, London: Macmillan.

Gosh, A. (1992), *In an Antique Land*, London: Granta Books.

Hirsch, E. and O'Hanlon, M. (eds) (1994), *The Anthropology of Landscape*, Oxford: Oxford University Press.

Küchler, S. (1993), 'Landscape as memory: the mapping of process and its representation in a Melanesian society', in B. Bender (ed), *Landscape: Politics and Perspectives*, Oxford: Berg.

La Violette, P. (forthcoming) in C. Tilley (ed.), *Material Culture and Social Identity*.

Mitchell, T. (1989), 'The world as exhibition', *Comparative Studies in Society and History* 31: 217–36.

Morphy, H. (1993), 'Colonialism, history and the construction of place: the politics of landscape in Northern Australia', in B. Bender (ed), *Landscape: Politics and Perspectives*, Oxford: Berg.

Pratt, M.L. (1985), 'Scratches on the face of the country; or, what Mr Barrow saw in the land of the bushmen', *Critical Inquiry* 12: 119–43.

Pred, A. (1984), 'Place as historically contingent process: structuration and the time-geography of becoming places', *Annals of the Association of American Geographers* 74(2): 279–97.

Said, E. (1989), 'Jane Austen and empire', in T. Eagleton (ed), *Raymond Williams: Critical Perspectives*, Oxford: Polity Press.

Selwyn, T. (1996), *The Tourist Image: Myths and Mythmaking in Tourism*, Chichester: John Wiley.

Strang, V. (2001), 'Negotiating the river: cultural tributaries in Far North Queensland', in B. Bender and M. Winer (ed.), *Contested Landscapes: Movement, Exile and Place*, Oxford: Berg.

Sutton, D. (1998), *Memories Cast in Stone*, Oxford: Berg.

Barbara Bender

Tilley, C. (1994), *A Phenomenology of Landscape*, Oxford: Berg.
—— (1999), *Metaphor and Material Culture*, Oxford: Blackwell.
Williams, R. (1973), *The Country and the City*, London: Chatto & Windus.

Contested Landscapes: Medieval to Present Day*

Barbara Bender

The Cumbales, who live in the Bolivian Andes, say:
Although events occurred in the past, we live their consequences today and must act upon them now. What has already occurred is in front of us, because that is where it can be corrected.

> J. Rappaport 'History and everyday life in the Colombian Andes'

Introduction

On a small, heavily populated, down-at-heel, offshore island, with illusions about its position in the world, the past can become oppressive. Fay Weldon, in her *Letter to Laura* (1984), describes what it is like to live in a country with too much past, too little present:

> Every acre of this tiny, densely populated land of ours has been observed, considered, valued, reckoned, pondered over, owned, bought, sold, hedged – and there's a dead man buried under every hedge, you know. He died of starvation, and his children too, because the common land was enclosed, hedged, taken from him . . . The past . . . is *serious*.

She is describing an English, rather than a Scottish or Welsh landscape, one in which, at least from medieval times, armed interventions were short-lived, where appropriations – and contestations – were somewhat more subtly negotiated. She gets the sense of the 'terrible beauty' to be found in many such landscapes, the tension between the pleasure gained from a worked-over, lived-in landscape and the uneasy knowledge of what the working and living often involved; the way the 'historical rootedness' of the English landscape, the seeming slow evolution, has served to disguise a proprietorial palimpsest, the working out of a long history of class relations.

Not just the land, but the very word 'landscape' has often been used as though it 'belongs' to a particular class. Recently British Rail put up two Intercity posters

* B. Bender, *Stonehenge: Making Space*, Oxford: Berg Publishers, 1998, pp. 97–131.

for First Class travel. One of them began with the words: 'The only Constable you'll see at a hundred miles an hour . . .' The assumption was clear: those who matter know who Constable is, those who might be confused by a seemingly obscure reference to the forces of law and order will not get the joke and do not matter. The other poster, reproduced in Fig. 6.1, shows a quiet landscape, monotone, almost monotonous, acceptable to those who know how (and can afford) to appreciate the understatement – and those desirous of joining their ranks. It is a view from a window. You are the observer. It is not your land but it is certainly *someone's* land. The fields are cultivated, but the cultivators and their machinery are not visible. You could perhaps view it as an 'old' landscape with the tree and the hedge standing for something stable and unchanging. Or you might view it as a 'new' landscape in which, in fact, most of the hedges have been grubbed up to make way for 'economies of scale' and agribusiness. There is perhaps, for those in the First Class, a satisfactory elision between old and new so that the changes associated with a monetarist climate are referenced on symbols of stability and the old order – the old tree – and thereby gain legitimacy.

In this chapter, working with one small (albeit heavily symbolic) corner of the English landscape, I want to explore not only the different ways in which, over a period of several hundred years, those with economic and political power and the necessary cultural capital have attempted, physically and aesthetically, to appropriate the landscape, but also how these appropriations have been contested by those engaging with the land in quite different ways. Elsewhere (Bender 1998, ch. 1), I have suggested that people's experience of the land is based on their everyday attentiveness to the tasks in hand, the routines, their relations to each other, to their animals and crops, and to the world around them, but that that engagement is also shaped by the particularity of the historical moment. The decoupling of 'the cultural' from the political, and the individual from the larger historical structures, found in recent postmodern writings, has to be resisted (Thrift 1991, Jackson 1991). We need to retain the coupling while recognizing the complexity of interactions, and eschewing any one-way causality. We may also, I believe, retain the notion of hegemonic discourse. There is, in fact far less difference between Gramsci and Foucault than people seem to think, and for the purposes of this chapter it is important to stress that those with power will attempt to impose particular ways of doing and seeing which serve to disguise the labour process. However, those with power, and the strategies they invoke, are not only riven by internal factions and tensions, but are dependent upon some degree of acceptance by those 'without' power, and must perforce take on board their 'reactions, contestations and subversions (Harvey 1996: 44, Keesing 1994, Thompson 1974, Williams 1994).

Having stressed that one can only understand the contestations and appropriations of a landscape by careful historical contextualization, it goes against the

Figure 6.1 Intercity travel poster

grain to attempt to sketch the history of the Stonehenge landscape over a period of several hundred years in one chapter. What follows can be no more than an outline and is piecemeal. I concentrate on the medieval, the landscapes of the seventeenth and eighteenth century, and, finally, in somewhat more detail, the present-day landscape.

For each period the evidence is very different. For the medieval period one can draw on both archaeological and literary evidence. But the latter is primarily derived from the work of clerics. There are references to the commoners, but only as the (despised) 'other'. As we move on in time, the description thickens, the voices quicken, but always, even in the present, there is unevenness in the way in which different people can be heard.

Medieval Stonehenge[1]

Anglo-Saxons had laid claim to the Celtic *Choir Gaur* and it had become Stan Hencg. Since the Saxons did not bother, on the whole, to rename natural features, this suggests that the site still carried meaning. Most probably the stones had been incorporated into folk culture, syncretic, changing, but possibly enduring over the millennia, or perhaps their significance had ebbed and flowed within people's consciousness (Piggott 1941).

We have the renaming, and then a long silence. The next mention of Stonehenge is in the later medieval period, in the middle of the twelfth century. And now the stones express the conflict between Church and commoner.

For most of the medieval period the Stonehenge downlands were marginal land, probably used for common grazing. To the east, the open fields belonging to the villages of Amesbury Countess in the north, and Amesbury West in the south straggled the valley bottom and lower hill-slopes of the river Avon. The stones lay on the edge of the familiar world; beyond lay the 'wilderness' (Anglo-Saxon *Wylder ness*, nest or lair of a wild beast) (Stilgoe 1982: 10). For the medieval commoner, the landscape, both familiar and unfamiliar, was imbued with magic – half-pagan, half-Christian. The stones in their liminal setting were revered and believed to have curative and procreative powers. In the villages, houses had to be protected from witches. Sometimes a carved rowen post would be set to one side of the hearth. Sometimes the post, carved with St Andrew crosses, was set in place by the local priest who, rarely a man of letters, trod a fine line between the old and the new religion (Stilgoe 1982). The higher echelons of the Church could protest. As early as the eighth century, King Canute exclaimed that:

> It is heathen practice if one worships idols, namely if one worships heathen gods and the sun, or the moon, fire or flood, wells or stones or any kind of forest tree, or if one practices witchcraft (cited in Burl 1979: 36).

But, as Le Goff makes clear, the Church authorities, faced with the power of passive resistance, negotiated, trimmed and adapted their doctrines (Gurevich 1988: 5, Le Goff 1980: 160-88). The commoner did likewise.

In many ways the positions of Church and commoner were not so far apart. For both, 'the material world was scarcely more than a sort of mask, behind which took place all the really important things . . . Nature . . . in the infinite detail of its illusory manifestations . . . conceived above all as the work of hidden wills' (Bloch 1962: 83). Both imbued landscape with magic, and there was no dearth of 'magical' Christian sites and relics. The problem was simply how to interpret the magic. Where the Church saw the hand of God, the commoners, in their fearful encounter with the wilderness, confused Satan with Pan, and continued to relate hoary tales of the wild hunt when, '. . . on moonless nights, and especially Walpurgis-nacht (May Day eve), Satan and his hounds coursed through the forest, pursuing with a terrible roaring and baying all the wild creatures and any humans unlucky enough to stumble in their way' (Stilgoe 1982: 8).

The Church, heir to the Graeco-Roman tradition, was more prone to define good and evil, true and false, black and white magic. The old religion – which was part old, part new – was more equivocal, more ambiguous: forces were good *and* bad, natural forces were to be propitiated and thereby rendered beneficial (Le Goff 1980: 159–88).

The Church, embedded in the hierarchical relations of a feudal society, preached of a 'natural order' homologous to feudalism; the pyramidal social relations were reiterated in the hierarchy of heavenly relations and in material representations. Demarcations – social, material, ritual – were to a large extent played out within a single body of space rather than, as later on, between distinctive, class-differentiated, spaces. The inside of a church, of a manor, even a poor man's house, was physically 'open', but socially and mentally heavily demarcated (Johnson 1993). In this world of interdependencies, the commoners, the worshippers of stones, were represented as 'the other' – their labour necessary, their visage hideous. They also acted as a foil for the qualities of wealthy men and saints, as 'fodder' for the redemption of the upper classes (Le Goff 1980: 95). There is little that is positive about the commoners in the writings of the clerics, and much that is passed over in silence. Our sense of their world is filtered through the contortions, concessions, and suppressions of the Church.

The Church physically 'appropriated' the stones. Amesbury, close to Stonehenge, housed a religious order,[2] and 'the presence of a religious house nearby may have been influential in bringing about the destruction of parts of the monument' (Cleal, Walker and Montague 1995: 343). At Avebury the Anglo-Saxon church was built alongside the great bank and ditch (but see Bender 1998: 139). The stones and mounds were also 'intellectually' appropriated and adulterated. They became a malignant part of a Christian iconography, the work of the Devil –

Devil's Den, Devil's Quoit, Devil's Chair. The great cone of Silbury Hill, near Avebury, became the place where the Devil had dropped a spadeful of earth. The Heel Stone at Stonehenge compounded a diabolical and Arthurian etymology. It bore, supposedly, the impression of a holy father's heel. The heel had been struck by a stone thrown by the Devil angered by Merlin, King Arthur's magician, magicking the stones into place (Burl 1979: 36).

The medieval Church vacillated between attempting to appropriate the stones, and destroying them. One of the earliest accounts of Stonehenge, c.1136, by the Welsh cleric Geoffrey of Monmouth, embedded in his *History of the Kings of Britain*, combines 'pure legend and a sense of the marvellous that is sometimes still completely pagan' (Bloch 1962: 100).

In Monmouth's account, Aeneas, arriving from Troy, conquered the giants that inhabited 'Albion'. Later, a Celtic King, Emrys (Ambrosius), brother to Ythr (Uther Pendragon, father of Arthur), intent on establishing a memorial to kinsmen killed at Amesbury by Saxon treachery, enlisted Merlin's help in transporting the great stones from Ireland and erecting them at Stonehenge (Fig. 6.2). According to Monmouth, these stones had originally been brought by giants from Africa to

Figure 6.2 Merlin hefting the stones into place (Egerton 3028 fl40v; kind permission of the British Library)

Ireland.[3] In a matter-of-fact way, he cites Merlin's description of the magical power of the stones:

> Whenever they felt ill, baths [w]ould be prepared at the foot of the stones; for they used to pour water over them and to run this water into baths in which their sick were cured. What is more, they mixed the water with herbal concoctions and so healed their wounds. There is not a single stone among them which hasn't some medicinal virtue (Geoffrey of Monmouth 1966: 196).

In later medieval times such benign empowering of the stones was forbidden and the Church moved to obliterate their magic.[4] At Avebury, in the early fourteenth century, the Church authorities made the local inhabitants destroy the stones. Nevertheless, the destruction was done with caution. The stones were not broken up; instead the villagers dug large holes and tumbled the stones into them, 'handled with reverence, covered without hurt to the sarsen, doing God's work without upsetting the Devil' (Burl 1979: 37). Compare this with the late seventeenth-century, when the local farmers broke up the stones, partly for purely utilitarian reasons, partly, as we shall see, as part of the Protestant backlash against paganism and popery. Now they were broken with fire and water, and were incorporated into local buildings (Fig. 6.3).

Figure 6.3 Breaking up the stones at Avebury (Gough Maps 231, fol 5r; kind permission of the Bodleian Library, Oxford)

The Seventeenth and Eighteenth Centuries

Ucko, Hunter, Clark and David. (1991: 163) suggest that by the end of the seventeenth-century the commoners no longer worshipped the stones and broke them with impunity. But perhaps this is too simple a reading. There remains a tension, right through the seventeenth and into the eighteenth century, between the stones seen as something utilitarian, a source of building material, an encumbrance to be got rid of by farmers 'chiefly out of covetousness of the little *area* of ground each stood on' (Stukeley, cited in Ucko et al. 1991: 249), and the stones invested, how-ever vestigially, with supernatural power. The Nonconformist preacher, harnessing all the populist rhetoric at his disposal to force a disassociation between 'pagan' ritual and festival and true Christian piety, not only succeeded in repudiating the 'emotional calendar of the poor', but also drove a wedge between the teachings of the Church and everyday life, between 'polite' and 'plebeian' culture. In attempting to destroy the 'bonds of idolatry and superstition – the wayside shrines, the gaudy church, the local miracle and cults' (Thompson 1974), he undermined traditional remedies against the devil and his agencies, and made witchcraft and pagan practices appear both more powerful and more menacing (Hill 1982). Added to which, by the end of the seventeenth century, with the acceleration of the enclosure movement, many a member of a church congregation had been cut loose not only from the land, but also – a negative freedom – from traditional forms of obligation and service (Thompson 1974). There was, thus, not only a degree of political anarchy, but also a considerable residue of superstition. Indeed it was not unknown, in the late seventeenth century, for preachers to attempt to 'tame' their congreg-ations by threats of 'petrification' for those who danced on the Sabbath, and several stone circles were attributed to just such divine intervention (Grinsell 1976). *Fools Bolt*, an anonymous diatribe also from the late seventeenth century, both threatens, and remains threatened by the power of the stones at Stonehenge:

> These forlorne Pillers of Stone are left to be our remembrancers, dissuading us from looking back in our hearts upon anything of Idolatry, and persuading us . . . so to . . . deride, it in it's uglie Coullers, that none of us . . . may returne, with Doggs, to such Vomit, or Sows to wallowing in such mire' (cited in Legg 1986: 4).[5]

There are many mentions of the magical properties of the stones in the seven-teenth and eighteenth centuries. John Aubrey, in the 1660s, notes, in the context of Stonehenge, that 'it is generally averred . . . that pieces of powder of these stones, putt into their wells, doe drive away the toades . . .' (cited in Olivier 1951: 157); while the Rev. James Brome noted in 1707: 'if the stones be rubbed, or scraped, and water thrown upon the scrapings, they will (some say) heal any green wound or old sore' (shades of Geoffrey of Monmouth nearly 400 years earlier) (cited in Burl 1987: 220). In the 1740s Stukeley records that people chipped off bits of bluestone because they were thought to have medicinal properties. And when John Wood, also in 1740, attempted to survey the stones and a violent storm blew up, the locals reckoned he had raised the devil (Burl 1987: 182).

In the longer term, the old ways, the old superstitions, the last vestiges of a 'deification of nature' were undermined by what Marx called 'the great civilizing influence of capital' (quoted in K. Thomas 1983: 23). The change in people's attitudes towards the natural world occurred in piecemeal fashion, depending upon who they were and where economically, politically, socially – they were located. An illiterate peasantry (slowly being commodified into farm-*hands*) abandoned more slowly the notion of 'a natural world redolent with human analogy and symbolic meaning, and sensitive to man's behaviour' (Thomas 1983: 89). It was the literate classes that embraced the concept of nature as something detached, 'to be viewed and studied by the observer from the outside, as if by peering through a window . . . a separate realm, offering no omens or signs, without human meaning and significance' (Thomas 1983: 89). And – more sinisterly – as something 'to be put to the question' (Bacon), racked for its secrets and treasures (Gold 1984).

The more pronounced attempts by the Church, in the seventeenth and eighteenth centuries, to impose Christian teachings and Christian marriage upon the vestigial paganism and easier-going sexual mores of the countryside, were only part of the greater intervention in the lives of ordinary people by both Church and State. Having followed through the contestation between Church and commoner, I have now to backtrack in order to discuss the way in which the State interceded, and to consider the importance of Stonehenge in the context of an emerging national, and then regional, sense of identity.

Already in the fifteenth century, the Church had begun to lose its exclusive hold on the past – history was no longer a matter of ecclesiastical precedent. 'Time' ceased to be the gift of God and became 'the property of man' (Le Goff 1980: 51), and 'Time's arrow' began to replace the repetitive cycles – the endless chain of cause and effect – of earlier history. Instead, historical precedents and 'genealogies' became part of the process of internal pacification and nation-state consolidation.

As early as the fourteenth century, Stonehenge begins to be drawn into a nationalist diatribe. Langtoft tells the story of 'the Wander Wit of Wiltshire', who:

> rambling to Rome to gaze at Antiquities, and there skewing himself into the company of Antiquarians, they entreated him to illustrate unto them that famous monument in the contry called Stonage. His answer was that he had never seen, scarce ever heard of it, whereupon they kicked him out of doors and bad him goe home and see Stoneage. And I wish that all such Episcopal cocks as slight these admired stones and scrape for barley cornes of vanity out of foreigne dunghills, might be handled, or rather footed, as he was (cited in Olivier 1951: 156).

Under the Tudors, historical and archaeological precedents were used to formalize custom and tradition into instruments of government and a defined code of law (Piggott 1985). And at a time when the Act of Union (1536) extended the English law of the land to Wales and menacingly set out the need 'utterly to extirpe all and singular the senister usages and customes differinge from [the Realme of Englande]' (cited in Jones 1990), the myth was created that the inhabitants of England and Wales were one people with a common ancestry and shared history (Jones 1990, Piggott 1985: 16). Henry VIII, intent on breaking with Rome and seizing the Church lands, employed Leland to ride around the country, mapping, collating, and 'spying'. And Leland, professing himself 'totally enflammed with a love to see thoroughly all partes of your opulente and ample realme . . . thereby to expose . . . the craftily coloured doctrine rout of Roman bishops', rode around the land:

> By the space of these vi yeares paste that there is almost nother cape nor bay, haven, creke, or peere, river or confluence of rivers, breches, washes, lakes, mere, fenny water, montaynes, valleis, mores, hethes, forestes, woodes, cities, burges, castelles, principal manor places, but I have seen them (cited in Chamberlin 1986: 69).

Inter alia, he mapped and described Stonehenge, appending, without comment, Geoffrey of Monmouth's explanation.

Compilations, categorizations and a comprehensive mapping of Britain became increasingly important during the seventeenth and early eighteenth centuries. Refinements in surveying and mapping were part of a changing technology of power, integral to the development of mercantile capital, and the opening up of the New World. These new techniques allowed the redefinition of property: 'It created geometrical, divisible, and hence saleable space by making parcels of property out of lands that had previously been defined according to rights of custom and demarcated by landmarks and topographic features' (Olwig 1996).

Control of knowledge, of resources and of nature. By the late seventeenth century, as the 'county naturalists picked their way through all the legends of prognosticatory springs, portentous birds and similar marvels', popular and learned views of nature significantly separated out (Thomas 1983: 78). James I demanded that Stonehenge be made to give up its secret and ordered his court architect, Inigo Jones, to map and explain it. Giants and Merlin were no longer considered sufficient. Inigo Jones, steeped in Italianate landscapes, claimed Stonehenge for the Romans. It was, he said, as he manipulated his plans, built c. AD 79 by British chieftains subject to Rome and was based on Vitruvian geometry. The king encouraged the Duke of Buckingham to dig the monument, causing – as, a little later, Aubrey laconically noted – the 'falling downe, or recumbancy of the great stone' (Fig. 6.4).

". . . So to make up for the disappointment
we had to let him have a home computer."

Figure 6.4 Reproduced by kind permission of Merrily Harpur

With the influx of capital into the countryside, with 'new' families competing with and marrying into the older aristocracy, and with both the old and new aristocracy hastening to enclose open field and common land in order to 'improve' their fortunes, the mapping and description of land-use, natural curiosities and the genealogies of local families helped both to create and legitimate the new class and property relations.

More and more, chorographers [created] books where country gentlemen [could] find their manors, monuments, and pedigrees copiously set forth. In just a few decades chorography thus progressed from being an adjunct to the chronicle of kings to become a topographically ordered set of real-estate and family chronicles (Helgerson 1986).

Aubrey, the 'discoverer' in 1666 of Avebury (the commoners had, of course, been there a while!) and of the Aubrey holes at Stonehenge, was the first to propose that Stonehenge was not Roman or Danish but rather Celtic and (probably) associated with the Druid religion. He was also not reluctant to admit that he was recording with the explicit intent that the knowledge be made available to those who might find economic advantage (McVicar 1984). Walter Charleton, three years earlier, was in favour of a Danish origin and proposed it as 'a *court royal*, or place for the *Election and Inauguration of Kings'*, and, in the same vein, dedicated it to Charles II (Ucko et al. 1991: 15). And John Smith of Boscombe, in the later eighteenth century, noting that Stonehenge was built 'to show the steady, uniform and orderly motions of the heavenly bodies', dedicated it to the Duke of Queensbury at nearby Amesbury House, 'as a symbol of your Grace's steady, uniform and orderly conduct through life' (Olivier 1951: 36). Based in large measure on the formidable treatise on *Brittania* by Camden (first published in 1586, republished 1695 in an enlarged edition), antiquarians increasingly focused on regional, rather than national, coverage.

The economic advantages reaped by the antiquarians' patrons were real enough. The commoners were physically evicted as part of the seventeenth – and eighteenth-century enclosure movement. They were also evicted aesthetically. In 1726, the Duke of Queensbury, owner of Amesbury House near to Stonehenge, enclosed the open fields of West Amesbury and Amesbury Countess to the east of Stonehenge, and laid great parts of the estate down to park. He realigned the Amesbury–Market Lavington road, landscaped the Iron Age camp and built a 'Druidical' grotto. In passing, he 'preserved' Stonehenge from a plague of rabbit warrens planted by his predecessor.

The antiquarian, William Stukeley, though a churchman and a friend of the gentry, occasionally hints at a darker local experience of landscape enclosure. Writing, in 1740, of his survey of the Stonehenge area, he says, 'this . . . will . . . preserve the memory of it hereafter, when the traces of this mighty work are obliterated by the plough . . . that instrument gaining ground too much, upon the ancient and innocent pastoritial (sic) life . . . and by destructive enclosures depopulates the country' (Royal Commission 1979: iv). It is, of course, the technology that Stukeley blames, rather than the changing social relations.

Stukeley, apart from his threnody about enclosure, was an Establishment man. In his writings on Stonehenge and Avebury there is a tension between his desire to record in meticulous 'scientific' detail and his need to validate his theological interpretation. There has been a tendency to periodicize Stukeley's life, separating out an earlier 'scientific' fieldwork phase from a later 'druidic' phase, and lamenting his lapse into romantic fabulation (Piggott 1985: 15). More recent work suggests that, from the beginning, the two went hand-in-hand (Ucko et al. 1991: 53), and now that we all admit to the subjectivity that imbues our practice, we can more readily accept the contradictions in Stukeley's work. So, on the one hand, in the early 1720s, he meticulously maps the cultural landscape of his forebears, engaging Stonehenge as part of a larger landscape. On the other, having read (and rather under-acknowledged) Aubrey, he promulgates a Druidic origin and, more, attempts to place the Druids in a direct line from Moses and Abraham – finding in the Druidic religion the lineaments of the Christian Trinity. Avebury becomes 'a landscaped model of the Trinity' in which the great circle represents 'the ineffable deity, the avenues . . . his son . . . in the form of a serpent'. Stonehenge, he believes, is a temple erected in 460 bc by Egyptian 'refugees', aided by Wessex Druids (Piggott 1985). While he held these notions from the 1720s and had 'revised' some of his plans accordingly, they were given added urgency in the 1730s and 1740s. Having moved from being a medical man to taking holy orders, perhaps in part because his hopes of patronage were disappointed, he found that Toland and the Deists were also promoting Stonehenge as a Druidical temple. Their intention was to relativize all religion and to suggest that 'Christianity was as old as the Creation'. In the face of this heretical intervention, Stukeley altered the title of his book from the original 'History of the Ancient Celts' to 'Patriarchal Christianity: or A Chronological History of the Origin and Progress of True Religion and Idolatry'.[6]

Stukeley appropriated the Druids in the service of the Church of England. Others, particularly the Welsh nationalists in the aftermath of the French Revolution, rediscovered them as early patriots, Celtic leaders of the opposition to the Romans, symbols of resistance and liberty (Jones 1990). In both Thomas's poem 'Liberty' and in Collins's 'Ode to Liberty' (1747) the Druids figure as the apostles of freedom (Piggott 1985).

Cleric, antiquarian, landed gentry, Welsh nationalist – the number of voices increases. Each appropriates Stonehenge in their own fashion, each creates a particular past. Some voices remain muted. The voice of the labourer, or of women of any class, most often come down to us at second-hand. Other people talk about them, or do not talk about them. Silences are important. And often the bitterness seeps through. The radical claim to the land that the Diggers and Levellers fought for in the seventeenth century was superseded, in the eighteenth century, by frustration and anger of the wholesale appropriation and enclosure of the land, as expressed, for example, in the writings of the poet-labourer John Clare (Bender 1998: 31).

In contemporary Britain, even with all the media available, it is still not easy to hear all the voices that contest the Stonehenge landscape. Moreover, while the stones remained 'open' right through to the beginning of this century and people could come to them with their different understandings, they are now 'closed' and Stonehenge has become a museum which attempts to 'sell', not always successfully, a particular sort of experience, a particular interpretation of the past. People with alternative views have to fight for the right of entry and the right to express their views.

Contemporary Landscapes

A Site for 'the Nation'

The first impulse to 'protect' Stonehenge emerged from late nineteenth-century radical protests at the effect of industrialization on people and places alike. On the one hand, 'romantic' socialists like John Ruskin, appalled by the destruction wrought by the Industrial Revolution, proposed in 1854 that an inventory of 'buildings of interest' threatened by demolition be drawn up. On the other hand, the fierce intercession by working-class socialists demanding access to the countryside for the urban working class led to the creation of the Youth Hostelling Association and the Ramblers Association (Bommes & Wright 1982). In the late 1870s, John Lubbock attempted to introduce a National Monuments Preservation Bill. Three times the bill was furiously opposed by conservative members of the House of Commons who recognized that public amenity might come to override the rights of private ownership (Murray 1989).[7] There was, for example, a fine intercession by Francis Hervey, a Tory M.P: 'Are the absurd relics of our barbarian predecessors,' he clearly roared, 'who found time hanging heavily on their hands, and set about piling up great barrows and rings of stones, to be preserved at the cost of the infringement of property rights?' (Bommes & Wright 1982). Tennyson once wrote a poem in which father and son are trotting across their estate. The sound of the hooves merge with the father's opening credo – 'Property, property, property'.

Eventually, following the precedent set by the Commons Preservation Society in the 1880s, a wonderful British blurring of the distinction between 'private' and 'public' ownership was concocted and the politicians were won over. The National Trust (founded in 1895) holds the properties and land privately in the national and public interest.

In 1894 Sir Edmund Antrobus, owner of Stonehenge, refused to allow the Ancient Monuments Commission to fence Stonehenge. He still saw it as an important public space. If they tried to fence the stones, he said, 'an indignant public might act as the London public did in regards to the railings of Hyde Park, when the claim to hold meetings was interfered with' (Legg 1986: 162). His son, however, offered Stonehenge to the nation, at the excessively high price of £125,000 and with the proviso that he retain hunting and grazing rights. When this offer was turned down, he threatened to sell it to the Americans. In 1901, with the approval of the Society of Antiquaries, he erected a barbed wire fence around the monument on the specious grounds that the new military camp on Salisbury Plain might result in damage. He put in two custodians and charged an entrance fee of a shilling a head (the equivalent of £5 today – obviously beyond the means of most people). Flinders Petrie and Lord Eversley (founder of the Commons Protection Society) took Antrobus to court, but as Eversley remarked, the judge appeared to regard with equanimity the exclusion from the monument of the great bulk of the public. He was evidently under the impression that the vulgar populace had, by their destructive propensities, disqualified themselves as visitors to a place of antiquarian interest (cited in Legg 1986: 1–6).

Eversley surmised that the judge was much influenced by a line from Horace which, freely translated, runs: 'I hate the profane crowd and I exclude them'. Ninety years later things have not changed that much.

In 1915 Stonehenge was put up for auction, and a Mr Chubb (perhaps ironically of the lock-and-key family) bought it for his wife after she remarked over breakfast that 'she would like to own it'. Mr Chubb paid £6,600. In 1918 he gave Stonehenge to the nation, with the express wish that access should remain free. Unfortunately he added the proviso 'unless the Ministry of Works deemed otherwise' (Legg 1986). The Establishment 'deemed otherwise', and Stonehenge became a museum. Fenced in, available only to those who pay, it is no longer part of a living landscape (Lowenthal 1979, but see Bender 1998: 140).

Protecting the Site 'for the Nation'

Jacquetta Hawkes's archaeological writings are no longer in vogue, but one terse statement has been cited and re-cited: 'Every age', she said, 'has the Stonehenge it deserves – or desires' (Hawkes 1967: 174). At the time, she was reacting to the commercial development of the Stonehenge visitors' centre. Her comment was apposite then, and remains so now. Only, now, the Stonehenge that we have has not only become a tacky tourist trap, but also, once a year, at the approach of the Summer Solstice, becomes a gulag. The arc lights go up, the razor wire unrolls, and police and security men patrol with their dogs. For a brief moment the physical force that sustains the power of the ruling classes visibly flexes its muscles. In 1985 five hundred 'travellers' (a label that covers a great assortment of people who move around the countryside in old vans and buses) or 'free festivalers' were arrested and two hundred vehicles were impounded. In 1988 there was a repeat perform- ance. In the last few years there has been an eerie silence: no one, except the police and selected journalists (and a few brave free festivalers crawling through the long grass), is there to watch the sun rise.

There is no doubt that the confrontational politics of the Thatcher years have thrown into high relief some of the conflicts that underwrite British society. It was not fortuitous that Wapping (the printers' strike), Orgreave (the miners' strike), and the Battle of the Beanfield at Stonehenge occurred within a year of each other. The same tactics, involving the deployment of non-local unidentifiable police, were used in each encounter. The same language defined and excommunicated 'the other'. For Mrs Thatcher the miners were 'the enemy within'. Assistant Chief

Figure 6.5 Reproduced by kind permission of Hector Breeze

Constable Clement admitted that he 'would not be the slightest bit troubled if . . . [the pickets] were trampled by horses'. As for the travellers, Douglas Hurd labelled them 'medieval brigands'; Mrs Thatcher brayed '[I will do] anything I can to make life difficult for such things as hippy convoys'; Sir John Cope, Tory MP, trumpeted 'we need a paramilitary police force'; and Robert Key, local Tory MP, gave away part of the reason for this anger and disquiet, 'two hundred nomads are squatting illegally on private land . . . there may be a sensible case for the use of troops.'

Figure 6.6 Reproduced by kind permission of Hector Breeze

Media coverage reflected, and continues to reflect, similar prejudices. The *News of the World* headline read 'Sex-mad junkie outlaws make the Hell's Angels look like little Noddy' (Rosenberger 1991). In a post-Thatcher, less confrontational mode, damages against the police have been awarded to both travellers and miners.[8] There have been vaguely conciliatory moves that never, however, materialize into anything solid. They muffle, rather than resolve, the underlying tensions.

The travellers, as an unpropertied, anarchic minority, enrage the Establishment. In particular, they enrage when they lay claim to Stonehenge. I shall return to them, but I want first to consider who attempts to call the tune. In an era of flexible capitalism, who are the power-brokers at Stonehenge? what are the economics of an imagined past? who writes the scripts? and who buys the product?

Who 'Owns' Stonehenge?

Who – officially – owns Stonehenge? English Heritage 'owns' the site 'for the nation'.[9] The National Trust 'owns' 1,500 acres around Stonehenge, also 'for the nation'. *Sotto voce*, the Ministry of Defence 'owns' vast stretches of Salisbury plain, presumably for the defence of the nation.[10] The new visitors' centre was scheduled to be built on the edge of their terrain, but this encroachment has been seen off.

English Heritage and the National Trust act in the interests of the – heterogeneous and tensioned – Establishment. Through the 1970s, 1980s and into the early 1990s the radicalism that once informed attempts to protect and open up the countryside evaporated. Radicalism was replaced by populist rhetoric – doing *battle* for *our* heritage, maintaining *communal* values (apolitical, organic, stable and deeply unequal), promoting *traditional* skills. A rhetoric that only thinly disguises both old and new class interests.[11]

The Stonehenge 'Honey-pot'

Contrary to a conservation rhetoric that emphasizes a 'green' small business economy and the encouragement of specialized skills, the reality of preservation, conservation and public access is very 'big' business. In a climate of industrial uncertainty and in the computerized age of flexible capitalism, investment in industry and property becomes more problematic. And the tourist market, which requires less investment in fixed facilities, looks increasingly attractive.[12] Although the £15 million (and rising) earmarked for the new Stonehenge visitors' centre seems a substantial amount, the returns will undoubtedly be equally substantial (Bender 1998: 181).[13]

In 1983 government responsibility for the preservation of ancient monuments and for scientific research was handed over to English Heritage – a quango, autonomous but State-funded. Lord Montagu of Beaulieu (author of *How to Live in a Stately Home and Make Money*) took the chair and one of the stated aims became the heavy promotion of tourism (Hewison 1987: ch.4).

Present-ing the Past

Stonehenge now has to make money. It also has to be preserved, conserved and presented. The visitors' centre – as everyone freely acknowledges – is a botched attempt to provide tourist facilities, keep people away from the archaeologically sensitive central part of the monument, and present information about the place and the landscape. It has been rightly berated for its neo-brutalist architecture, inadequate parking and facile 'macho' historical presentation.

Until very recently, the tunnel that went under the road took one back through time via an Astronaut, Henry VIII, the Roman Empire, Egypt and the Pharaohs – not a woman in sight – to Stonehenge and the Beginning of Civilization. The time-tunnel debouched onto the green sward surrounding the stones. There were (again until recently – they have now been moved to the car-park) three dioramas. The first depicts a sort of rural arcadia, a pastoral scene with early Stonehenge – the bank and ditch – just visible in the background (Fig. 6.7). In the second, a man stands poised between rural arcadia and civilization (Fig. 6.8). In the third a chief – modelled on Kirk Douglas – brandishes his insignia of power as he talks to his male cronies (Fig. 6.9). In the background the labour force beavers away. The emphasis is on power and on technology – building, construction, planning, control.

It is a very particular sort of history. One that goes hand in hand with the more general conservative version (vision?) of the past offered by English Heritage:

> The National Heritage is remarkably broad and rich . . . it is simultaneously a representation of the development of aesthetic expression and a testimony to the role played by the nation in world history (The First Annual Report of the National Heritage Memorial Fund, cited in Hewison 1987).

Apart from acting as the custodians and entrepreneurs of prehistoric monuments, English Heritage and the National Trust have focused on the landmarks of those with power and wealth, inscribed in an aesthetic that bypasses, as it has done for centuries, the labour that created the wealth. More recently the net has been cast wider, but, despite acquiring Victorian back-to-backs or derelict mills or mines, the presentation remains sanitized and romanticized, emphasizing local colour rather than the socio-economic conditions that generate both wealth and poverty, people's pain or their resistance.

Figure 6.7 English Heritage reconstruction of the first phase at Stonehenge (photo: B. Bender)

Figure 6.8 Reconstruction of phase 2 at Stonehenge (photo: B. Bender)

Figure 6.9 Reconstruction of phase 3 at Stonehenge (photo: B. Bender)

Stonehenge is 'explained' in terms of roots, and of 'our' 'deep' national past. It tells the story of those empowered to make decisions and to make claims on people's labour. It is a top-down past that ends when the last stone was put in place. It is a museum exhibit, a place to be looked at, caught on film, consumed on site and, via photographs and souvenirs (and memories), back home. Hence the incongruity of the marginalized travellers or Druids laying claim to the place and perceiving it as a 'living' space, a meeting-place, a ritual centre.

Stonehenge becomes 'our' national icon, exemplar of past glories, part of our national identity, part of our justification for remaining – against the odds – in the top league of world players. And yet, at the same time, and whilst being vigorously reworked as a commercial proposition, it exemplifies 'England's green and pleasant land', bulwark against the forces of modernity.

And the Role of the Archaeologists

The conservation and marketing lobby depend for their information and explanation of the past on the academic establishment – in the case of Stonehenge, the archaeologists. Archaeologists justify their monopoly on information in terms of the scientific rigour of their discipline. They have the artefacts, the dating materials and methods that permit the rigorous reconstruction of the past. In reality the archaeological excavations at Stonehenge throughout this century have been piecemeal, often slipshod and, until recently, unpublished, and archaeologists still understand relatively little about the social and political conditions and the cultural perceptions of the people who built and used Stonehenge.[14] Nonetheless, as the 'official' interpreters, archaeologists, by and large, repudiate the 'alternative' theories of the New Agers, and side with the conservationists in limiting access to the stones and preventing celebrations on the grounds of potential damage. Only one academic book – edited by Chippindale – has been published that presents the attitudes and aspirations of a wider range of people, and even then there is no attempt to suggest possible compromises (Chippindale et al. 1990). When the police dug a trench 15 feet long and 6 feet wide across the entrance to the Free Festival field the archaeologists failed to protest, just as they have never protested at the despoliation of the landscape to the north of Stonehenge by the Ministry of Defence firing ranges and tank runs. Archaeologists attempt to eschew politics. One of the archaeologists on the Commission that repudiated the Free Festival, washed his hands of the predictable confrontation that would occur. 'It is most important,' he said, 'to draw the distinction between the defensible decisions as such and their executive consequences, largely in the hands of others' (Fowler 1990). To the travellers the archaeologists are the 'unconscious apologists for industrial civilisation', and the looters of graves (Thompson cited in Chippindale 1983: 248).[15]

Figure 6.10 Reproduced by kind permission of Merrily Harpur

Who Gets to Go?

For whom is the site preserved and the explanations offered? Mainly for the tourists who pay their money and are then corralled through the barriers and along the pathways roped off from the centre. Only those with academic credentials, or those in the advertising industry with enough money to pay for privileged access (Fig. 6.11), cross the ropes and, under strict supervision, enter the stones. Those who do not have money, or are unwilling to pay, are kept out. By force, if necessary (Fig. 6.12).

Bona fide tourists are the ones who have, since 1901, paid their entrance fee. In the 1920s there were 20,000 paying visitors; in 1955 184,000; in 1977 815,000; in 1982 530,000; in 1994 672,000. On a hot summer's day in the peak year of 1997, 7000 people arrived in the course of a day and there were 2,000 visitors in one hour. Nearly three-quarters come from overseas, and of those nearly half are from the United States. Only 18 per cent of the visitors thought Stonehenge was good value for money! (Golding 1989).

Tourism at Stonehenge as elsewhere, is an expression of the easefulness of long-distance communication and travel. 'Why', as Jencks put it, 'if one can afford to live in different ages and cultures, restrict oneself to the present, the local? Eclecticism is the natural evolution of a culture with choice' (cited in Harvey 1989: 87).

For many people 'sightseeing', as Bourdieu (1984) pointed out, is a form of symbolic capital: putting the emphasis on knowing where to go and what to see disguises both the surplus wealth required and the way in which symbolic capital

Figure 6.11 Consuming the stones

Figure 6.12 'Protecting ' the stones (photo: Alan Lodge)

translates into 'real' capital, increasing the 'value' of those who possess it. But 'sight-seeing' – connecting up with the past (multiple fragmented pasts), creating memories, reminiscing – is also part of the way in which people construct a sense of identity. And different sorts of tourists construct different sorts of identity (Urry 1990), and for some the significance of their experiences may well be 'deeper and be more complex than conventional studies and surveys of leisure and tourism . . . imply' (Clark et al. 1994). Those who come by car, bicycle or on foot distinguish themselves from the 'mass' tourists who arrive by tour-bus with a 'dwell-time' of twenty-five minutes (more than half of which, it has been ascertained, is taken up with buying souvenirs, food and going to the toilets). They may prefer to see themselves as 'travellers' (hippy or high-class). Those who come by bus will have equally diverse agendas, and diverse perceptions of what they are doing and seeing. It is a mistake to assume that people are the passive recipients or dupes of the heritage industry. Just as Hall (1980) and others have shown with popular culture, so with heritage, people mould and reconstruct their experiences. As Warren (1993) puts it, 'Speaking the language of fantasy, [tourists] can remain sublimely outside conventional structures of logic and always just beyond the reach of the dominant hegemonic forces'. People's understandings of the places they visit are wound around with memories, resonances and unpredictable connections. Reality, Lacan once suggested, is merely a prop on which we lean our dreams (as cited in Crang 1994). And in the 'epoch of juxtaposition, the epoch of the near and far, of

" We decided to come after all to see some of your traditional hippy convoys "

Figure 6.13 Reproduced by kind permission of Hector Breeze

the side-by-side, of the dispersed' (Foucault 1986: 22), the 'views' of the 'foot-loose' traveller/tourist will be impressionistic and comparative, and resonate curiously and variably with the heritage packaging of a deep-rooted and stable past.[16]

And Who Doesn't

So many different interest groups, so many different understandings of the Stonehenge landscapes. Finally we have the 'alternative' landscapes of both the eccentric but respectable Druids – people who often hold down perfectly good jobs but reject orthodox religion – and the travellers and free festivalers, who are often more sweeping in their rejection of Establishment orthodoxies.

The Druids take their cue from Stukeley. The ancient Order of Druids was founded in 1781, as part of the Romantic movement, happy to pick up where Stukeley had left off. A splinter group, the United Order of Druids, was founded in 1833. The Druids worship at the Summer Solstice but they only started coming to Stonehenge in 1905. Up to the 1970s the Druids were tolerated by the Establishment. They were 'exotic', and quite good for the tourist trade (Fig. 6.15).

The unacceptable 'weirdos' are the New Age travellers and myriad other groupings (hippies, punks, bikers, musicians, clowns, jugglers, peace activists, Hell's Angels, Quakers, Hare Krishna devotees) who held their Free Festival in the field next to Stonehenge for a decade, from the mid-1970s, culminating in the 1984 Festival which attracted over 50,000 people (Rosenberger 1991) (Fig. 6.17). For

Figure 6.14 Reproduced by kind permission of *London Evening Standard*

Figure 6.15 Druidic celebration at Stonehenge (photo: Alan Lodge)

FESTIVAL EYE   SUMMER 1989

Figure 6.16 Reproduced by kind permission of Pete Loveday

many of them, Stonehenge is (was) an important meeting-place, a place for spiritual and other sorts of celebrations, weddings, exchanges, part of a seasonal circuit of summer festivals, winter park-ups on commons or derelict urban sites. Many of them, like the Druids, believe that there are psychic forces at Stonehenge – energy fields, leylines – or that it is a temple for the worship of the sun and the moon, for the renewal of seasons.[17]

One free festivaler said:

> It's the main part of the earth energy system . . . It's like a *chakra* on the earth's energy. You've got your acupuncture – the system of the human body with all the meridians and all the burial mounds marking them. Then you've got the *chakras* of the human body which are the bigger ones like Old Jerusalem, Mecca, Egypt, Easter Island and Stonehenge . . . (Willie X: personal communication)

For several years the authorities remained tolerant. After 1978 the stones were roped off, and standpipes, temporary lavatories and rubbish collection points were installed in the Festival field (Golding 1989). There was very little vandalism, and there was considerable self-policing. But by the early 1980s government and media had turned against the travellers' self-named Peace Convoy, fearing them as anarchists and connecting them with the politically unpopular Peace Camps at Greenham Common, Molesworth, and elsewhere.[18] Despite the 1968 Caravan Sites

Figure 6.17 Free Festival at Stonehenge (photo: Alan Lodge)

Act which requires local authorities to provide sites for travellers, such places became fewer and fewer. In 1985 the National Trust and English Heritage took out an injunction and the police moved against them (Chippindale 1986). There were violent showdowns in 1985 and 1988.

The police have spent over £5 million policing Stonehenge. The government have passed a Public Order Act and a Criminal Justice Act. The police can now arrest two or more people 'unlawfully proceeding in a given direction', and can create 'exclusion zones' to prevent confrontation. The antagonism towards the traveller is not surprising. At the end of the day England's landscape is a proprietorial palimpsest. The travellers own no land or houses, and pay no direct taxes.

Conclusion

What I have tried to do is to chart, and to begin to explain, a multitude of voices and landscapes through time, mobilizing different histories, differently empowered, fragmented, but explicable within the historical particularity of British social and economic relations, and a larger global economy. Marxist theory is not in fashion, but, as Harvey notes, re-negotiated it still resonates to our condition: 'What Marx depicts . . . are social processes at work under capitalism conducive to individualism, alienation, ephemerality, innovation, creative destruction,

"When two or more are gathered
together you're in breach of
the Public Order Act".

Figure 6.18

Figure 6.19

speculative development – a shifting experience of space and time, as well as a crisis-ridden dynamic of social change' (Harvey 1989: 111). Does that not begin to engage with the confrontations – emotive, intellectual and physical – that surround contemporary Stonehenge? Marx's 'social processes' were of course focused along class lines. They must be widened and cross-cut by gender, age, ethnicity and so on (Deutsche 1991, Gilroy 1987, Massey 1991). And too, the power of rhetoric, and the lived materiality of our existence – of which the time

Figure 6.20 Reproduced by kind permission of Steve Bell

tunnel, tarmac pathways and the ropes are only small indicators – must be recognized and included as integral to these social processes.

If this chapter is, towards the end, somewhat polemic, that is because it was, in part, spawned in anger at the efforts of English Heritage and parts of the Establishment to promote a socially empty view of the past in line with modern conservative sensibilities. I hope that it begins to justify the study of landscape, not as an aesthetic, not as grist for the First Class Intercity poster, but as something political, dynamic and contested.

Notes

1. In this medieval section, there are many allusions to Avebury, the other great stone circle that lies 30 km. to the north. Avebury is less well known than Stonehenge, though for some, like John Aubrey, it 'did as much excell *Stoneheng*, as a cathedral does a Parish church' (cited in Hunter 1975: 158). A medieval village straddled part of the Avebury circle, and the church, built sometime after 900 AD, was located just beyond the bank and ditch. Because of the proximity of the church and village to the stones, there is often more detailed information about Avebury than about Stonehenge. Obviously there were differences between these medieval landscapes, but also much that was similar.

2. There is some suggestion of an early minster church, superseded by an abbey in 979 AD, and in turn replaced by a priory of nuns in 1177 (Cleal et al. 1995: 343).

3. Piggott (1941), much impressed by this part of Monmouth's saga, suggests that it indicates that Monmouth was heir to the Welsh bardic tradition that, in faint outline resonated with millennia-old traditions recounting the movement of the bluestones from Wales to Wessex.

4. In a contemporary setting, Herzfeld comments on the need to destroy the past. In Crete, during the militantly nationalist regime of 1967–74, 'one official had even wanted to demolish the proud minaret at the mosque of *gazi* Huseyin Pasa. As long as it stood, he reasoned, the Turks could use this monument to make territorial claims on Crete. History had to be remodeled to make the present safe' (Herzfeld 1991: 57).

5. Various guesses at the authorship had been made: J. Gibbons perhaps (Piggott 1985: 86), or Robert Gay, a Parliamentarian rector (Legg 1986). Whoever he was, the author was exceedingly scornful of Inigo Jones – 'Out-I-goe Jones' – and of his Romano-British origin for Stonehenge.

6. The first entry to the Index of the volume, published in 1740, runs:
They were of the patriarchal religion Page 1, 2, 17
Which was the same as Christianity 2, 54

7. The political implications of celebrating a Celtic past at a time when the English were putting down the Irish was not lost on the House of Commons (Murray 1989).

8. The 'Battle of the Beanfield' took place in June 1985. Six years later, in June 1991, £23,000 was awarded to 24 plaintiffs for 'assault, damage to their vehicles and property, and for not being given the reasons for their arrest'. However, the judge managed to find the police not guilty of unlawful arrest and they therefore did not have to pay the costs of the trial. So the plaintiffs' awards were swallowed up in legal costs. The miners' confrontation with the police (or vice-versa) at Orgreave in Yorkshire occurred in June 1984. In July 1985 the case against the 95 pickets was dropped. In June 1991 39 (former) miners shared £425,000 out-of-court compensation from the South Yorkshire police for assault, wrongful arrest, malicious prosecution and false imprisonment (*Guardian*, 20 June 1991).

9. More accurately, the land on which Stonehenge stands is owned by the Department of the Environment, but its management is entrusted to English Heritage, a quango established by Act of Parliament in 1983 (Golding 1989).

10. Between 1897 and 1902 the War Office negotiated the purchase of 43,000 acres to the North of Stonehenge. Subsequent purchases have brought it up to 91,000 acres (36,400 ha) (Cleal et al. 1995: 346).

11. Jacobs (1994a, 1994b, 1996: 41, 43, 102) discusses this rhetoric in the context of the 'SAVE Britain's Heritage' campaign mounted in the mid-1970s in the face of threats by the then Labour government to impose a Wealth Tax detrimental to the propertied classes. The Victoria and Albert Museum put on an exhibition on the threat to country houses which seeded this campaign.

12. Although the figures are somewhat out of date, Lumley (1988) estimated the cost of creating a job in manufacturing at £32,000, in mechanical engineering at £300,000, and in tourism at £4,000.

13. The takings from the Yorvik (Viking) shop at York were, per square foot, and prior to recession, more than every Marks and Spencers in the land except the Oxford Street branch (Baker 1988). The number of museums increased by over 50 per cent during the 1970s and 1980s. Roughly one museum opened every three weeks and by the late 1980s there were over 2,000 museums in Great Britain (Hewison 1987: ch.1). With the recession, some have proved short-lived.

14. Ascherson (1988) remarked on two current renditions of the transition from Late Neolithic to Early Bronze Age. The first draws a contrast between those whose power derives from communal ritual expressed in collective architectural enterprises, and a new breed of thrusting individuals whose richly equipped single inhumations show a contempt both for traditional aristocratic ritual and for collective values. In the alternative reading, the Old Neolithic establishment were 'conservatives laden with gold and divine knowledge, scheming to defend their influence against local rebels and sceptics – who, in turn, chafed to get rid of the old frauds and run things in a modern way with Beaker pottery and metal tools'. Yet another version turns Stonehenge into an 'astronomical laboratory' – Fred Hoyle, Astronomer Royal, even suggested that the astronomers were, perhaps, a genetically distinct group (Chippindale 1983: 264). As Hilary, Wes and Paul (Bender 1998: 90) point out, present-day social configurations often seem to weigh rather heavily in our interpretations of the past.

15. Sir Edmond Antrobus (who protested against the fencing of Stonehenge) grumbled in the House of Commons in 1874 that 'some of the ancient barrows, through having been first rifled by antiquarians, have been carted away and levelled by farmers. For himself, he believed it was the antiquarians who had done the most mischief in England' (cited in Murray 1989: 63). He wasn't far wrong: in the first decade of the nineteenth century William Cunnington and Sir Richard Colt Hoare 'ravaged' over three hundred barrows (Richards 1991: 33).

16. Urry suggests that with the increasing flow of capital and people across national borders, people may lose a sense of a coherent national culture. They by-pass the national to create a sense of local identity, whilst at the same time garnering a more global sense of the past through fragmented TV presentations and tourist forays (Urry, workshop, National Trust Centenary Conference, Manchester 1995).

17. On the other hand, the messages are not always so spiritual. In 1989 four great letters were grafittied onto the stone: L – I – V – E. They have been cleaned

off, of course, but no doubt they will still be visible under X-ray. Were they a proclamation about the life-force, or a political credo, or – as rumoured – the unfinished logo of a certain football team?

18. V.S. Naipaul in *The Enigma of Arrival*, was not enthused by the travellers: 'not gypsies . . . but young city people, some of them criminals, who moved about Wiltshire and Somerset, in old cars and vans and caravans, looking for festivals, communities, camp sites . . . As a deterrent Mr Phillips had the round building wound about with barbed wire' (Naipaul 1987: 270).

References

Ascherson, N. (1998), *Games with Shadows*, London: Radius/Century Hutchinson.

Baker, F. (1988), 'Archaeology and the heritage industry', *Archaeological Review from Cambridge* 7(2): 141–44.

Bender, B. (1998), *Stonehenge: Making Space*, Oxford: Berg.

Bloch, M. (1962), *Feudal Society*, L. Manyon, (tr.), vol. 1. London: Routledge & Kegan Paul.

Bommes, M. and Wright, P. (1982), 'Charms of residence, the public and the past', in R. Johnson et al. (eds), *Making Histories: Studies in History Writing and Politics*, London: Hutchinson.

Bourdieu, P. (1984), *Distinction*, London: Routledge & Kegan Paul.

Burl, A. (1979), *Prehistoric Avebury*, New Haven: Yale University Press.

—— (1987), *The Stonehenge People*, London: J.M. Dent.

Chamberlin, R. (1986), *The Idea of England*, London: Thames & Hudson.

Chippindale, C. (1983), *Stonehenge Complete*, London: Thames & Hudson.

—— (1986), 'Stonehenge: events and issues at the Summer solstice', *World Archaeology* 18(1): 38–58.

—— et al. (1990), *Who Owns Stonehenge?* London: Batsford.

Clark, G. et al. (1994), *Leisure Landscapes*, Lancaster: Centre for the Study of Environmental Change.

Cleal, R., Walker, K. and Montague, R. (1995), *Stonehenge in its Landscape. Twentieth-century Excavations*, London: English Heritage.

Deutsche, R. (1991), 'Boys town', *Environment and Planning D: Society and Space* 9: 5–30.

Fowler, P. (1990), 'Academic claims and responsibilities', in C. Chippindale et al. (eds), *Who Owns Stonehenge?* London; Batsford.

Geoffrey de Monmouth (1966), *The History of the Kings of Britain*, L. Thorpe (tr.) Harmondsworth: Penguin.

Gilroy, P. (1987), *There Ain't No Black in the Union Jack: the Cultural Politics of Race and Nation*, London: Hutchinson.

Gold, M. (1984), 'A history of nature', in D. Massey and J. Allen (eds, *Geography Matters*, London: Macmillan.

Golding, F. (1989), 'Stonehenge – past and future', in H. Cleere (ed.), *Archaeological Heritage Management in the Modern World*, London: Unwin Hyman.

Grinsell, L. (1976), *Folklore of Prehistoric Sites in Britain*, Newton Abbot: David & Charles.

Gurevich, A. (1988), *Medieval Popular Culture: Problems of Belief and Perception*, Cambridge: Cambridge University Press.

Harvey, D. (1989), *The Condition of Postmodernity*, Cambridge: Blackwell.

—— (1996), *Justice, Nature and the Geography of Difference*, Cambridge, MA: Blackwell Publishers Inc.

Hawkes, J. (1967), 'God in the machine', *Antiquity* 41: 174–80.

Helgerson, R. (1986), 'The land speaks: cartography, chorography, and subversion in Renaissance England', *Representations* 16: 51–85.

Herzfeld, M. (1991), *A Place in History. Social and Monumental Time in a Cretan Town*, Princeton: Princeton University Press.

Hewison, R. (1987), *The Heritage Industry: Britain in a Climate of Decline*, London: Methuen.

Hill, C. (1982), 'Science and magic in seventeenth-century England', in R. Samuel and G. Stedman Jones (eds), *Culture, Ideology and Politics*, London: Routledge & Kegan Paul.

Hunter, M. (1975), *John Aubrey and the Realm of Learning*, London: Duckworth.

Jackson, P. (1991), 'Repositioning social and cultural geography', in C. Philo (compiler), *Reconceptualising Social and Cultural Geography*, Aberystwyth: Cambrian Printers.

Jacobs, J. (1994a), 'Negotiating the heart: heritage, development and identity in postimperial London', *Environment and Planning D: Society and Space* 12: 751–72.

—— (1994b), 'The Battle of Bank Junction: the contested iconography of capital', in S. Corbridge, R. Martin and N. Thrift (eds), *Money, Power, Space*, Oxford: Blackwell.

—— (1996), *Edge of Empire. Postcolonialism and the City*, London: Routledge.

Johnson, M. (1993), 'Notes towards and archaeology of capitalism', in C. Tilley (ed.), *Interpreting Archaeology*, Oxford: Berg.

Jones, R. (1990), 'Sylwadau cynfrodor ar Gôr y Cewri; or a British aboriginal's land claim to Stonehenge', in C. Chippindale et al. (eds), *Who Owns Stonehenge?* London: Batsford.

Keesing, R. (1994), 'Colonial and counter-colonial discourse in Melanesia', *Critique of Anthropology* 14(1): 41–58.

Le Goff, J. (1980), *Time, Work and Culture in the Middle Ages*, A. Goldhammer, (tr.), Chicago: University of Chicago Press.

Legg, R. (1986), *Stonehenge Antiquaries*, Sherborne: Dorset Publishing Co.

Massey, D. (1991), 'Flexible sexism', *Environment and Planning D: Society and Space* 9: 31–57.

McVicar, J. (1984), 'Change and the growth of antiquarian studies in Tudor and Stuart England', *Archaeological Review from Cambridge* 3(1): 48–67.

Murray, T. (1989), 'The history, philosophy and sociology of archaeology: the case of the Ancient Monuments Protection Act (1882)', in V. Pinsky and A. Wylie (eds), *Critical Traditions in Contemporary Archaeology*, Cambridge: Cambridge University Press.

Naipaul, V.S. (1987), *The Enigma of Arrival*, Harmondsworth: Penguin.

Olivier, E. (1951), *Wiltshire*, London: Robert Hale.

Olwig, K. (1996), 'Recovering the substantive nature of landscape', *Annals of the Association of American Geographers* 86: 630–53.

Piggott, S. (1941), 'The sources of Geoffrey of Monmouth. II. The Stonehenge story', *Antiquity* 15: 305–19.

—— (1985), *William Stukeley. An Eighteenth Century Antiquarian*, London: Thames & Hudson.

Rappaport, J. (1988), 'History and everyday life in the Colombian Andes', *Man* 23: 718–39.

Richards, J. (1991), *Stonehenge*, London: Batsford/English Heritage.

Rosenberger, A. (1991), 'Stones that cry out', *The Guardian* June.

Royal Commission on Historical Monuments (1979), *Stonehenge and its Environs. Monuments and Landscape*, Edinburgh: Edinburgh University Press.

Stilgoe, J. (1982), *Common Landscape of America 1580 to 1845*, New Haven: Yale University Press.

Thomas, K. (1983), *Man and the Natural World*, Harmondsworth: Penguin.

Thompson, E. (1974), 'Patrician society, Plebian culture', *Journal of Social History* 7: 382–405.

Thrift, N. (1991), 'Over-wordy worlds? Thoughts and worries', in C. Philo (compiler), *Reconceptualising Social and Cultural Geography*, Aberystwyth: Cambrian Printers.

Ucko, P., Hunter, M., Clark, A. and David, A. (1991), *Avebury Reconsidered: from the 1660s to the 1990s*, London: Unwin Hyman.

Urry, J. (1990), *The Tourist Gaze*, London: Sage Publications.

Weldon, F. (1984), 'Letter to Laura', in R. Mabey, S. Clifford and., A. King (eds), *Second Nature*, London: Jonathan Cape.

Williams, R. (1994), Selections from 'Marxism and Literature', in N. Dirks, G. Eley and S. Ortner (eds), *Culture/Power/History: a Reader in Contemporary Social Theory*, Princeton: Princeton University Press.

−7−

Memory and Conflict

Nicholas J. Saunders

Introduction

Anthropologists have long been interested in war, though hitherto attention has focused mainly either on conflicts among tribal peoples and chiefdoms, or been concerned primarily with trade, politics and the origins of the state (e.g. Haas 1990, Fried and Murphy 1968). To date, the study of twentieth century conflict in particular has relied overwhelmingly on military history's accounts of events (e.g. Gilbert 1989, Keegan 1998), and the economic, social and political consequences of individual wars (e.g. Barnett 1987, Ferguson 1998). Apart from an art historical interest in war painting (e.g. Cork 1994), and a broader concern with post-conflict commemorative monuments (e.g. al-Khalil 1991, King 1998, Rowlands 2001), the audits of war reveal an anthropological focus on the materiality of conflict per se to have been virtually absent.

Yet, as first-hand memory of the twentieth century's major conflicts fades, our view of these events is increasingly determined by interpretations of material culture by those alienated from its production and original purpose. Here, the multi-disciplinary approach of material culture studies comes into its own, offering new ways of investigating industrialized war at a personal and cultural level and on a regional as well as global scale. The fact that modern conflicts are defined by their technologies as wars of *matériel*, is a clear invitation for such an approach.

In recent years, several explicitly historical works on war have transcended their disciplinary boundaries and appeared increasingly, if unintentionally, anthropological (e.g. Bourke 1996, Keegan 1996, Winter 1995). Nevertheless, the pace of recent developments in interdisciplinary academic endeavours and in society's attitudes to war is such that there is an urgent need for an explicitly anthropological/archaeological approach to modern warfare. In particular, what is required is a re-evaluation of the role of material culture as multi-vocal representational embodiments of war (Schofield et al 2002, Saunders n.d.).

Materialities of Conflict

War is the transformation of matter through the agency of destruction, and industrialized conflict creates and destroys on a larger scale than at any time in human history. Modern war has an unprecedented capacity to remake individuals, cities, nations and continents. The immense production of material culture during industrialized conflict and the extremes of human behaviours which it embodies and provokes, suggests that in a very real sense we can begin to talk about the 'cosmology of war'.

An anthropological focus on material culture sees objects as possessing important and variable social dimensions beyond, as well as including, their original design purpose. Objects embody an individual's experiences and attitudes as well as cultural choices in the technology of production. They occupy a dynamic point of interplay between animate and inanimate worlds, inviting us to look beyond the physical world and consider the hybrid (and constantly renegotiated) relationships between objects and people (Attfield 2000: 1).

War objects may be small, e.g. a bullet; intermediate, e.g. a tank; or large, e.g. a whole battlefield landscape. All share the defining characteristic of being the product of human action rather than natural processes. Thus, the Western Front of the First World War is as much an artefact as a portable war souvenir, a Second World War V2 rocket, the symbolic terrain of war memorials, or the 'Cross' formed by remaining structural elements in the ruins of the World Trade Centre.

Seeing the world in this way, identifying and engaging with artefacts of all scales, allows us to construct a biography of the object (*pace* Kopytoff 1986) – to explore its 'social life' by assessing the changing values and attitudes attached to it by different people over time (*pace* Appadurai 1986). The features of war, like any artefacts, embody a diversity – but perhaps a unique intensity – of individual, social and cultural ideas and experiences. Their analysis reveals the social origin of artefact variability (Miller 1985: 1), and the fact that at the same time they are part of, and constitute the physical world. This world structures perceptions, constraining or unleashing ideas and emotions by the people who live within it (Miller 1985: 204–5). Illustrating this point is the change in attitudes in Britain towards the commemorative association of war memorials, Armistice Day, and the observance of two minutes silence from 1919 to the present (Gregory 1994, Richardson 1998). Here, materiality, spirituality, politics and emotion link the living with the dead in a complex interplay of past and present.

Consciously and subconsciously, we all interact with the endlessly varied objects that surround us. The constant intimacy of people with countless things – our immersion in the material medium – may well be the most distinctive and significant feature of human life (Schiffer 1999: 2,4). In wartime, this medium is constituted more intensely than in peacetime, and human interactions with it (and each other) reflect this intensity. If we accept that an individual's social being is determined by their relationship to the objects that represent them, that objects become metaphors for the self and thereby constitute a way of knowing oneself through things (Hoskins 1998: 195), then we may also consider that objects make people as much as people make objects (Miller 1998, Pels 1998). Arguably nowhere is this more true than in modern war.

Objects, Scale and Intersection

The transformational power of industrialized conflict is evident at every scale of human activity, and can be tracked through a variety of materialities and across disciplinary boundaries. The investigation of trench art, for example, illustrates how the anthropological analysis of one corpus of war-related objects can initiate a new kind of debate on the nature of war. In the eclectic spirit of a material culture studies approach, it shows how such objects represent '. . . the visible knot which ties together an invisible skein of relations, fanning out into social space and social time' (Gell 1998: 62). In its varied shapes and kinds, trench art radiates a sense of inextricability – a conceptual coherence that challenges us to follow and understand as it migrates from one form and context to another.

What, we might ask, are the implications for assessing the 'social life' of First World War memorabilia displayed as 'memory objects' in the home for eighty years, identical items stored and/or exhibited in museums, and similarly identical materials excavated from Great War archaeological sites – some of which then feed and stimulate the international trade in military collectables? Here, attention is focused on a set of issues which have hardly been recognized by the relevant disciplines of social anthropology, museology and archaeology, let alone problematized or investigated.

For example, trench art has hitherto been ignored by military, cultural and art historians both as an artistic genre and as a unique war-related phenomenon. This despite the fact that every soldier and family of the war generation was familiar with these objects, and that Renoir, Rodin and Lalique presided over wartime exhibitions of these objects in Paris (ENOA 1915). Ignored by art history, these items nevertheless illustrate, in sometimes poignant fashion, the point made by Gell (1998: 74) in his attempt to create a true anthropology of art – i.e. how the artistic elaboration of artefacts attach people to things and to the social projects those things entail.

The promiscuity of meaning which characterizes these objects includes a role as ambiguous souvenirs in the post-war phenomenon of battlefield pilgrimage and tourism. Significantly, this role envelops the small 'memory object' within the larger one that is the palimpsest of multivocal landscapes of the Great War (Saunders 2001, Winter 1995). Objects and landscapes merge in the individual's experience of 'being in' a place (Tilley 1994) that is both real and imagined. In its modern variant, the itineraries of battlefield tours are commercially edited perceptions of reality, punctuated by stops where tourists eat lunch, drink coffee and buy souvenirs. During the inter-war years, many such items were bought by the bereaved, transported from the landscape of battle to the domesticated social space of the home where they were displayed, mediating grief and loss in ways which have been neither acknowledged nor documented. In both cases, object/landscape experiences create liasons which punctuate the textual dimension of memory. The investigation of these issues cuts across, and suggests potential collaborative projects between, the disciplines of anthropology, war studies, cultural history, tourism studies and cultural geography to mention a few.

However, the relationship between 'memory object' and 'memory landscape' takes on added significance as the investigation of landscape has also become a central concern of anthropological archaeology in recent years (e.g. Bender and Winer 2001, Tilley 1994, Ucko and Layton 1999). While the embedding of the object in a landscape can be an explicitly archaeological event, until recently there has been not even the pretence of a scientific archaeology of the Great War (Saunders 2002), or any other twentieth century conflict (see Schofield et al. 2002, Cockcroft 2001). Neither has there been a mention of archaeology in recent assessments of war and memory (e.g. Ashplant et al. 2000).

The technical, ethical and political challenges of creating such an archaeology are significant, as they cut across issues such as the excavation of still lethal battlefields and the recovery, identification and re-burial of the multi-faith and multi-ethnic dead. Such an archaeology also must acknowledge the need for constructing method-ologies for coping with the sensitive management of battle-zone landscapes as national and trans-national cultural heritage locations and tourist destinations. In Europe, these concerns extend from the First World War of 1914–18 to the Bosnian conflict of 1992–5, and intersect the materialities of religious belief (e.g. Becker 1998), and forensic investigations associated with ethnic cleansing and genocide. More widely, an archaeology of twentieth century conflict can be considered and theorized as but one of the many and appositely termed 'archaeologies of the contemporary past' (Buchli and Lucas 2001).

The material culture of war – particularly small and portable souvenirs, trophies, jewellery and equipment – intersects directly with modern archaeological practice and discourse. Due to their personalized nature, such objects are often attached – literally and figuratively – to bodies and body parts, and which, sometimes exclusively, are able to identify human remains and permit official and familial closure when reburied. It is these same items which are increasingly sought by militaria collectors who acquire them from illegal covert digging. Here, archaeological process must acknowledge an anthropological dimension, as a soldier's name, identity and place in society – i.e. his reclamation from the list of 'The Missing' – are rendered impossible through subordin-ation to commercial imperatives. These imperatives forever seal 'social being' within the object, alienating it from its rightful owner.

The complex relationships and associations of the material culture of twentieth century war are often characterized by seemingly endless ironies which loop back upon themselves and open up previously unrecognized areas for investigation, particularly by the varied and innovative anthropological approaches of those working within a material culture studies milieu.

Many objects are made from recycled war *materiél* which, in its original form as ordnance, helped create the battlefield landscape now turned archaeological site. The memory of these ambiguous events of destruction/creation can be recalled from the constituent parts of the object whose original design purpose (i.e. killing and maiming) is still evident. The classification of these kinds of objects by anthropology as recyclia (Saunders 2000, and see Cerny and Seriff 1996) connects the archaeological excavation of artefacts with wider notions of recycled and rejuvenated landscapes, associated monuments and reconfigured identities and spiritualities (e.g. Mosse 1991).

Arguably more than any other kind of cultural matter, the objects of war provide opportunities for exploring the ways in which the dead and the living find proximity via materialities and places (Hallam and Hockey 2001: 6). In the cultural life of objects, human beings are defined by their technologies, and nowhere are these technologies more insistent than in modern wars and their seemingly interminable aftermaths. As anthropology and archaeology begin to engage with these issues, the paradigm-shaping implications of an interdisciplinary approach to conflict are becoming clear.

References

al-Khalil, S. (1991), *The Monument: Art, vulgarity and Responsibility in Iraq*, London: Andre Deutsch.

Appadurai, A. (1986), 'Introduction: Commodities and the Politics of Value', in, A. Appadurai (ed.), *The Social Life of Things*, Cambridge: Cambridge University Press, 3–63.

Ashplant, T.G., Dawson, G. and Roper, M. (eds) (2000),. *The Politics of War and Commemoration*, London: Routledge.

Attfield, J. (2000),. *Wild Things: The Material Culture of Everyday Life*, Oxford: Berg.

Barnett, C. (1987), *The Audit of War*, London: Papermac.

Becker, A. (1998), *War and Faith*, Oxford: Berg.

Bender, B. and Winer, M. (eds) (2001) *Contested Landscapes*: *Landscapes of Movement and Exile*, Oxford: Berg.

Bourke, J. (1996), *Dismembering the Male: Men's Bodies, Britain and the Great War*, London: Reaktion Books.

Buchli, V. and Lucas, G. (eds) (2001), *Archaeologies of the Contemporary Past*, London: Routledge.

Cerny, C. and Seriff, S. (eds) (1996), *Recycled Re-seen: Folk Art from the Global Scrap Heap*, New York: Harry N. Abrams and Museum of New Mexico.

Cocroft, W.D. (2001), *Dangerous Energy: The Archaeology of Gunpowder and Military Explosives Manufacture*, London: English Heritage.

Cork, R. (1994), *A Bitter Truth: Avant-Garde Art and the Great War*, New Haven: Yale University Press.

ENOA. (1915), *Exposition Nationale des Oeuvres des Artistes tués a l'ennemi, Bléssés, Prisonniers, et aux Armées*, Catalogue to Exhibition, Salles du Jeu de Paume, Tuileries, Paris, May 20–July 20, 1915. Organized by 'La Triennale'.

Ferguson, N. (1998), *The Pity of War*, London: Allen Lane.

Fried, M. and Murphy, R. (eds) (1968), *War: The Anthropology of Social Conflict and Aggression*, New York: Doubleday.

Gell, A. (1998), *Art and Agency*, Oxford: Oxford University Press.

Gilbert, M. (1989), *Second World War*, London: Weidenfeld and Nicolson.

Gregory, A. (1994), *The Silence of Memory: Armistice Day 1919–1946*, Oxford: Berg.

Haas, J. (ed.) (1990), *The Anthropology of War*, Cambridge: Cambridge University Press.

Hallam, E. and Hockey, J. (2001), *Death, Memory and Material Culture*, Oxford: Berg.

Hoskins, J. (1998), *Biographical Objects: How Things Tell the Stories of People's Lives*, London: Routledge.

Keegan, J. (1996), *The Face of Battle*, London: Pimlico.

—— (1998), *The First World War*, London: Hutchinson.

King, A. (1998), *Memorials of the Great War in Britain: The Symbolism and Politics of Remembrance*, Oxford: Berg.

Kopytoff, I. (1986), 'The Cultural Biography of Things: Commoditization as Process', in A. Appadurai (ed.), *The Social Life of Things*, Cambridge: Cambridge University Press, 64–91.

Miller, D. (1985), *Artefacts as Categories*. Cambridge: Cambridge University Press.

—— (1998), 'Introduction', in D. Miller (ed.), *Material Cultures: Why some things matter*, London: UCL Press, 3–21.

Mosse, G.L. (1991), *Fallen Soldiers: Reshaping the Memory of the World War*, Oxford: Oxford University Press.

Pels, P. (1998), 'The Spirit of Matter: On Fetish, Rarity, Fact, and Fancy', in, P. Spyer (ed.), *Border Fetishisms: Material Objects in Unstable Spaces*, London: Routledge, 91–121.

Richardson, M. (1998), 'A Changing Meaning for Armistice Day', in H. Cecil and P.H. Liddle (eds), *At The Eleventh Hour: Reflections, Hopes and Anxieties at the Closing of the Great War, 1918*, Barnsley: Pen and Sword, 347–56.

Rowlands, M. (2001), 'Remembering to Forget: Sublimation as Sacrifice in War Memorials', in A. Forty and S. Küchler (eds), *The Art of Forgetting*, Oxford: Berg, 129–46.

Saunders, N.J. (2000), 'Trench Art: The Recyclia of War', in J. Coote, C. Morton, and J. Nicholson (eds), *Transformations: The Art of Recycling*, Oxford: Pitt Rivers Museum, 64–7.

—— (2001), 'Matter and memory in the landscapes of conflict: The Western Front 1914–1999', in B. Bender and M. Winer (eds), *Contested Landscapes: Movement, Exile and Place*, Oxford: Berg, 37–53.

—— (2002), 'Excavating Memories: Archaeology and the Great War, 1914–2001, *Antiquity* 76(1): 101–8.

—— (n.d.), *Materialities of Conflict: The Great War, 1914–2001*, London: Routledge. (In Preparation)

Schiffer, M.B. (1999), *The Material Life of Human Beings: Artifacts, Behaviour, and Communication*, London: Routledge.

Schofield, J., Johnson, W.G. and Beck, C. (eds) (2001), *Matériel Culture: The Archaeology of 20th Century Conflict*, London: Routledge.

Tilley, C. (1994), *A Phenomenology of Landscape*, Oxford: Berg.

Ucko, P.J. and Layton, R. (eds) (1999), *The Archaeology of Landscape: Shaping your Landscape*, London: Routledge.

Winter, J. (1995), *Sites of Memory, Sites of Mourning: The Great War in European Cultural History*, Cambridge: Cambridge University Press.

Bodies of Metal, Shells of Memory: 'Trench Art' and the Great War Re-cycled*

Nicholas J. Saunders

Sit on the bed. I'm blind, and three parts shell.
Be careful; can't shake hands now; never shall.

Wilfred Owen, *A Terre*

Materialschlacht and Material Culture

At the close of the twentieth century, the Great War of 1914–18 stands at the furthest edge of living memory. As the last survivors pass away, the past of 80 years ago survives increasingly as interpretations of material culture. History becomes archaeology, inviting, perhaps demanding, a new approach to the study of the world's first industrialized war.

Since 1914, Great War studies have focused on the analysis of military, economic, social and political histories, sometimes textured by memoirs and poetry. Almost totally absent has been any consideration of the physicality of what – even 84 years ago – was realized to be 'the war of *matériel*', *materialschlacht* (Terraine 1996: 11).

The First World War offers significant opportunities for a material culture approach. There is evidence already of broader, anthropologically inflected enquiries (e.g. Bourke 1996, Derez 1997, Liddle and Richardson 1997, Winter 1995, and see Kelly 1998). Nevertheless, there exists little appreciation of the interconnectedness and potential of archaeology and anthropology to inform analysis of the Great War – or of the ability of material objects to act as a bridge between mental and physical worlds (Miller 1987: 99).

Apart from annual Armistice Day ceremonies, where two minutes' silence symbolically compresses the memory of all war dead, the Great War survives – in material form and to a considerable degree – through the existence of and trade in a category of artefacts known collectively as 'Trench Art' – a misleading catch-all term applied to objects of various materials – made by soldiers and civilians alike between 1914 and 1939. The voluminous literature on the Great War carries hardly

* *Journal of Material Culture*, vol. 5(1), 2000, pp. 43–67.

a mention of these objects known to all soldiers and their families during the war and inter-war years. Partly as a consequence, they have remained unstudied for over 80 years.

Trench Art

Definitions of Trench Art are notoriously vague, the term itself an evocative misnomer. Where the National Army Museum refers to it as 'Decorative Arts', the Royal Air Force Museum labels it 'Commemorative Art', the National Maritime Museum keeps it in the Antiquities Department, and the Imperial War Museum allocates it to its Department of Exhibits and Firearms. A survey of over 100 regimental museums throughout the United Kingdom reveals similar ambiguity, with curators recognizing the term 'Trench Art' while questioning its validity and definition (Saunders n.d.a).

The problems of terminology are clearly linked to constituency. One well-known book on military collectables regards Trench Art as First and Second World War objects of metal, cloth, bone and wood made as souvenirs both during and after these conflicts (Lyndhurst 1983: 182–5). Strictly speaking the net should be cast more widely, to include objects made by, for example – Napoleonic prisoners of war, soldiers in the Franco-Prussian and Boer wars, as well as those who fought in, or were affected by, every twentieth-century conflict (including Bosnia-Hercegovina – see Saunders 2000c, Saunders and Wenzel n.d.). Any attempt to order this material requires a working definition. That adopted here is: 'any item made by soldiers, prisoners of war, and civilians, from war *matériel* directly, or any other material, *as long as it and they are associated temporally and/or spatially with armed conflict or its consequences*'.

To avoid problems raised by such inclusivity, and yet open a meaningful debate, this article maintains a narrow focus – the unique corpus of metal forms produced during the Great War and inter-war years along the Western Front and which gave its name to the genre. Specifically, items made from shells, shell cases, detonators, bullets, grenades, shrapnel, numerous ship parts, aircraft wire and engine parts, and a host of miscellaneous metal scrap. Apart from introducing metal Trench Art to material culture studies (*pace* Walters 1997: 63), I have three main purposes here: to bring a degree of order to its heterogeneous mass of objects, to explore the social lives of these artefacts, and, more widely, to illustrate the potential of a material culture approach to the study of the Great War – indeed, all war (Saunders 2000a).

From a material culture perspective (Editorial 1996; and see V&A/RCA 1998, 1999), the study of Trench Art intersects many issues which concern archaeology and anthropology. Trench Art pieces are objectifications of the self, symbolizing grief, loss and mourning (e.g. Maas and Dietrich 1994); are poignantly associated with memory and landscape; and with issues of heritage, and museum displays which increasingly emphasize the common soldier's experience of war. They are associated also with pilgrimage and tourism – particularly as regards their symbolic status as souvenirs – within which field their worked forms are ambiguously situated alongside the 'raw' unaltered mementoes of war (see Kwint 1998: 261, Richardson 1996). In addition, such objects are a prime example of recyclia (Saunders 2000b, and see Cerny and Seriff 1996). In short, metal Trench Art is an embodiment of the complex relationship between human beings and the things they make, use and recycle – in the physical, spiritual and metaphorical worlds they construct and inhabit.

Hitherto, 'Great War Art' has referred usually to paintings (e.g. Gough 1997, Harries and Harries 1983, Holmes 1918, and see Cork 1994), and also to architectural memorials erected at home and abroad, mainly between 1919 and 1937 (e.g. Boorman 1988, Borg, 1991). Although the grim, mechanical, anti-human aspects of the war were reflected in such art as *La Guerre and Cannon in Action* by the painter Gino Severini (Silver 1989: 75–6, Figs 41,42), the war was seen by many artists as a valueless and formless experience which could not be rendered by the conventions of the day (Hynes 1990: 108, and see Nash 1998: 29–30). Artists and writers approached it as archaeologists sorting through some gigantic midden the 'heap of broken images' of which Eliot's poem *The Waste Land* is constituted (Hynes 1990: 394–5).

Paintings and memorials represented war from a distance, spatially and temporally. They connected through impressions, possessing little or no sensuous or tactile immediacy. By contrast, metal Trench Art was made from the waste of war, its varied forms incorporating the agents of death and mutilation directly. Anonymously responsible for untold suffering and bereavement, expended shells, bullets and shrapnel were worked into a variety of forms, engaging visual, olfactory, tactile, and sometimes auditory senses, as well as memory. For civilian buyers, these objects also engaged the senses (though differently), often becoming an integral part of the 'house-world' of the owner, articulating the ritualized habits of domestic space in a complex working-out of personal emotions in the inter-war years.

Rich in symbolism and irony, metal Trench Art is a complex kind of material culture, whose physicality and nature make it a unique mediator between men and women, soldier and civilian, individual and industrialized society, the nations which fought the war, and, perhaps most of all, between the living and the dead.

Fragments as Categories

Trench Art today appears a vaguely defined body of 'lumpen ephemera' whose classificatory limbo enables it to be regarded and sold as antiques, militaria, bric-a-brac, 'awkward' museum pieces, curios and souvenirs. Behind this facade of ambiguity, a tangled mass of symbolism reveals changing social attitudes and cultural valuations linked indissolubly to the Great War and its aftermath.

Hitherto, the sheer quantity and diversity of this material appears to have discouraged any attempt at description or analysis. Nevertheless, it is possible to disentangle some of the strands and to identify a number of categories each with its own spatial, temporal and physical features, and symbolic associations. The meanings of these categories lie not only in variable form and use, but also in the trajectories the objects have taken (Appadurai 1986: 6), and in what Pels (1998: 94) has called the 'spirit of matter' – the ability of material objects to 'speak' and 'act' on their own (as well as through a multitude of interpreters).

Metal Trench Art is a prime example of the social nature of artefact variability (Miller 1985: 1) – its different kinds reflecting changes in British (and wider European) society in the wake of war, serving to reproduce the society (or that part of it) which consumed them. In the quarter century between 1914 and 1939, the various kinds of metal Trench Art made by the war generation – together with memorials, cenotaphs and remembrance ceremonies (see King 1998) – became part of the succeeding generation's given environment, structuring their perceptions and constructing them as subjects. In this sense, as Miller (1985: 204–5) says, 'objects do not merely constitute a given world which can be thus manipulated, but . . . these same objectifying processes also have consequences for the creation of the artefactual world itself'. Metal Trench Art helps us to understand a world which it simultaneously constituted.

Although it is possible to provide 'cultural biographies' (*pace* Kopytoff 1986) for several broad categories, it is unlikely that all items will be identified or classified. There are simply too many which do not fit, or which transgress neat boundaries to appear in more than one category. Ubiquitous metal matchbox covers, for example, were made by front-line soldiers, service personnel in rear areas, local civilians both during and after the conflict, and prisoners of war (e.g. LC: PNM 26, POW 9). While similarity of form cannot be taken to imply convergence of meaning, it appears that classificatory 'untidyness' is an integral and significant part of the nature of this material. The three categories formulated below are an initial attempt at ordering a vast array of objects, and are based on a limited though broadly representative number of examples.

Part 1: Categories of Trench Art

Category 1: 1914–19

Trench Art made by soldiers – in the front-line or rear areas – is the smallest category numerically speaking as its manufacture was restricted to the period of war and demobilization. Nevertheless, it displays the greatest variety of forms.

While many soldiers carved in chalk, wood or bone in the trenches, conditions under fire were always thought to have precluded manufacture of anything other than the crudest metal objects. This assumption concealed a far more interesting truth. Many sophisticated items were indeed made in 'view of the enemy'.

In the early years of the war, French and Belgian soldiers made finger rings from the aluminium parts of incoming German shells. These were melted down, poured into a mould, then filed, engraved and polished (R.W. 1915; and see Becker 1998: 100). Particularly revealing is the account of a British soldier who whiled away the hours in the trenches by buying 'a transfer from a Belgian soldier for five woodbines . . . then transferred the design to a shell with a bent nail . . .' In this way he decorated two 18-pounder shell cases with 'nouveau style female figures and flowers', inscribing one 'Souvenir of Loos', the other 'Souvenir of Ypres' (IWM, 131/89 CUP 5 Shelf I). On returning home, he polished and lacquered them, keeping both on his mantelpiece for 60 years.

The bulk of this category however was made in safer rear areas (Fig. 7.1) by off-duty soldiers, and especially farriers, Royal Engineers and service battalions such as the Chinese Labour Corps. The latter made it their business to find out which regiments were in the area, and then made metal Trench Art items decorated with appropriate badges and buttons to sell as souvenirs (Angela Kelsall, personal communication 1998); The Royal Engineers were similarly opportunistic as

Figure 7.1 Belgian soldiers decorating artillery shell cases north of Ypres B1186 147, Collections Musée royal de l'Armée Bruxelles

revealed by an extraordinary cache of half-finished and completed bullet letter-openers and bullets decorated with badges from various regiments stored inside two polished and lidded 4.5-inch howitzer cartridge cases (LC: S22; and see Anon. 1998: colour illustration).

Typical examples of category 1 items include:

1. Cigarette lighters made from bullets.
2. Matchbox covers made from brass or steel scrap (Fig. 7.2).
3. Letter openers made from bullets and scrap, sometimes inscribed, and often with badges attached.
4. Tobacco boxes and cigarette cases.
5. Military caps made from the base of shells.
6. Pens made from bullet cartridge cases.
7. Finger rings made from aluminium or brass.
8. Miscellaneous personal items decorated with, or made from, bullets or shell fragments.
9. Decorated artillery shell cases.

Apart from the obvious functionality of many such items, more elaborate pieces were also produced: an officer's 'swagger stick' artfully made from the bullets, in the Imperial War Museum (Saunders and Cornish n.d.); and miniature biplanes produced from bullet cartridges and metal scrap (e.g. Lyndhurst 1983: 182–3).

Figure 7.2 Brass matchbox cover, decorated with two halves of a 303 bullet, and a button. Photo: N.J. Saunders

While many objects were made 'on spec', others were clearly made to order, engraved with a man's name, and occasionally rank and regiment. This personalization has led many relatives to mistakenly attribute manufacture directly to a grandfather or uncle (Jane Peek, personal communication 1998). It is clear that most examples were made by individuals who had the time, safe location, expertise and access to tools with which to cut, smooth, shape, weld, solder and engrave metal.

Category 1 ceased to be made in the period between the Armistice of November 1918 and the signing of the Peace Treaty of Versailles in July 1919, during which time the majority of servicemen were demobilized and returned home.

Category 2: 1914–39

Economic deprivation and the incredible melange of available war debris produced a thriving civilian industry in metal Trench Art in France and Belgium for 25 years. Partly as a consequence, category 2, subdivided into a and b sections, was by far the largest quantitatively speaking. Items were mainly ornamental rather than functional, and the variety of shapes fewer than category 1.

Differences between sub-categories 2a and 2b are important yet paradoxically often difficult to establish. In both cases, identical forms often were made by the same people with the same techniques. With the exception of dated pieces, and those whose inscriptions implied a finished war, differences were less dependent on form and substance than on changing relations of production and consumption associated with the temporal shift from war to peace. While 2a was sold to Allied and German soldiers during the war, 2b was sold to war widows, pilgrims, and battlefield tourists between 1919 and 1939.

Typical examples of category 2 items include:

1. Brass shell cases, sometimes shaped, often decorated with floral designs (Fig. 7.3), and frequently engraved with the name of a town and/or region, a date, and such inscriptions as 'Souvenir of the Great War'.
2. Ashtrays made from or decorated with shell cases and bullet cartridges, sometimes inscribed, and often more elaborate than category 1 examples.
3. Letter openers, often inscribed and sometimes more elaborate than category 1 examples (Fig. 7.4).
4. Bullet-crucifixes made of cartridges with Christ figures or regimental badges attached.
5. Small decorated shell cases often mounted on a tripod of British, German or French bullets.

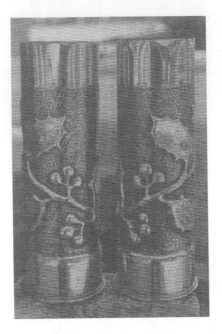

Figure 7.3 Pair of British 18 pounder brass artillery shell cases, one dated 1917, the other 1918, and decorated with 'holly and berry' design. Photo: N.J. Saunders

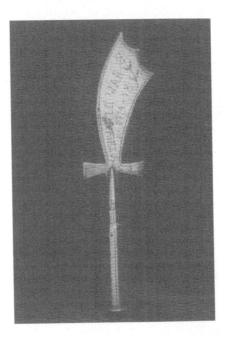

Figure 7.4 Brass letter-opener, made from a 1918 303 bullet. Photo: N.J. Saunders

Sub-category 2a: 1914–18. In a war-ravaged economy, civilian manufacture of metal Trench Art quickly developed into a cottage industry. Allied and German armies were large, if shifting, markets for such items. French and Belgian civilians often found themselves caught on different sides of the front-line, making Trench Art which could be sold as fortune decreed. For example, brass matchbox covers exist which depict the typical spiked German *picklehaube* helmet on one side, the inscription 'Gott mit uns' on the other, and 'Fabrique en France' inscribed along the spine (Ken Dunn, personal communication 1998).

A unique and intriguing aspect of this sub-category centres on the wartime legal status of expended British shells. Officially, these were not scrap, and, wherever possible, were collected into dumps, returned to Britain and refilled (Fig. 7.5). Using them to make Trench Art was technically illegal. In practice, different nationalities and the quantities of shells involved meant there was always a ready supply of such raw materials.

Sub-category 2b: 1919–39. After the Armistice, harsh economic realities persisted along the Western Front. Towns were devastated and a primarily agricultural landscape rendered useless (and dangerous) by saturation shelling. In the Ypres salient, up to five unexploded shells could be found in one square metre, and some 5,000 kg of shrapnel and detonators per hectare (Derez 1997: 443). Collecting and

Figure 7.5 Soldiers resting on a dump of artillery shell cases fired during the Battle of Polygon Wood, on the Ypres-Menin Road, Belgium, 30 September 1917. Photograph courtesy of Imperial War Museum, London

neutralizing this was no longer stealing but patriotic duty in advance of recon-struction (Clout 1996, and see Winter 1979: 263). Yet, an ever increasing quantity of raw material with which to make Trench Art was matched by a correspondingly rapid decrease of soldiers to buy it.

Between 1919 and 1939 however, a new market appeared in the large (albeit fluctuating) numbers of pilgrims and tourists visiting the battlefields and their associated memorials and cemeteries (Lloyd 1994). Ironically, those who once sold metal Trench Art to British soldiers now sold often identical items to the bereaved widows, sweethearts and relatives of servicemen who had not returned. More ironic still, some of the bereaved must have bought souvenirs fashioned from the very shells and bullets which they themselves had produced in munitions factories during the war.

In 1939, the advent of the Second World War stopped the flow of pilgrims and visitors, and the market for category 2b items abruptly ended.

Category 3: c. 1918–c. 39.

Made in Britain (and possibly elsewhere) mainly after the Armistice this is the most clearly defined category, temporally, spatially, in its distinctive forms, and in the identity of its manufacturers. Objects were fashioned from the 'raw' unworked materials of war brought back as souvenirs by returning service personnel, rather than the (usually) anonymous finished items of category 2. Manufacture was a commercial undertaking by various British firms such as the Army and Navy Store (Fig. 7.6), who offered to personalize soldiers' memorabilia by creating distinctive designs and mounting them typically on an ebonized base.

Often more elaborate than category 2 items, these mounted (and thus essentially 'civilized') forms were sometimes referred to as 'Mounted War Trophies'. They clearly owed much to British imperial traditions of displaying shields-and-spears, lion heads and tiger skins, from encounters in the far flung corners of empire.

Typical examples of category 3 items include:

1. Clocks made from shells and bullets.
2. Lamps and candlesticks from shells and bullets.
3. Inkwells, made from grenades and/or shrapnel.
4. Simple mounted shrapnel fragments
5. Various size 'cups made from shell parts
6. 'Table gongs' made from different size shell cases suspended from an ornate frame.

Almost all were designed for the domestic peacetime lives of returning soldiers, taming the experience of war, softening harsh memories and embodying the

Figure 7.6 Army and Navy Store advertisement for the mounting of 'War Trophies'

'swords into ploughshares' philosophy – a phrase sometimes engraved on such pieces. They probably functioned as visual reminders of wartime experiences – a bullet or piece of shrapnel which had wounded but not killed, a reminder of an all too close brush with eternity. The manufacture of this category seems to have ceased by the beginning of the Second World War, though possibly earlier.

Part 2: Trench Art Biography and Memory

What emerges from this assessment is the existence of different kinds of metal Trench Art – each resonating in distinctive ways with those who came (and continue to come) into contact with it. Meanings were not fixed in wartime or afterwards, and a single object could elicit a variety of responses. I will now explore some of these responses together with the wider connections and deeper issues associated with these artefacts.

Memories of Metal and Landscape

The emotive, enduring, and largely misleading term Trench Art originated with category 1, and is forever linked to the resonances these objects had for soldiers. Perhaps nowhere is the object's capacity to embody soldiers' experiences of war more apparent than in the account by Sapper Stanley K. Pearl (Australian 5th Field Company Engineers) of his making a Trench Art clock. This rare 'unpacking' of details offers a glimpse of the wealth of information and associations now largely lost. The piece was

> Made at Ypres in March 1918. The case was made from two 4.5-inch shell-cases picked up on Christmas Day 1917 at the Australian batteries at Le Bizet. The foot support is a clip of an 18-pounder shell. The arms are detonator wells of rifle-grenades and nose-caps. The hands are from a gun-cotton case, while the alarm cover is an American-made 18-pounder nose-cap with a 'whizz-bang' driving-band. The Rising Sun is the badge of a mate killed at Noreuil, while a button from the maker's greatcoat and a German bullet surmount the whole. (AWM, 14155)

Exposed to hitherto unimaginable quantities of bombs, mortars, shrapnel and bullets, servicemen like Sapper Pearl developed an ambiguous relationship with the metals of war – hardly surprising when almost three-quarters of wounds sustained were shell wounds (Winter 1979: 117). The confusions of battle and its aftermath mixed metal and flesh in a grisly mosaic. As Sergeant H.E. May observed in the Ypres salient in 1917, such a scene

> . . . was a vision indescribable in its naked horror. Pieces of metal that once were cannon; and, if good Krupp steel had been so shattered, what of the humans who served the steel? Heads, legs, arms, trunks, pieces of rotting flesh, skulls that grinned hideously, bones cleaned by exposure, lay about in hopeless riot. (May 1997: 200)

The impossibility of reconstituting men's bodies was in stark and ironic contrast to the creation of metal Trench Art from the scattered remains of shells, bullets, and shrapnel.

The intensity of physical and psychological experiences endured in battle led not only to shell-shock (see Simpson 1996), but also, possibly, to stripping away the atomizing effects of modernity, and inducing synaesthesic experiences in some individuals. In other words, destroying the dominance of the visual (especially in trenches and dugouts), and reinstating the olfactory, auditory and tactile elements of sensory experience (see Eksteins 1990: 146,150–1; and Howes 1991: 3–5). In this way, perhaps, new meanings were ascribed to the lights, sounds, smells, tastes and vibrations of war on such a massive scale – and which are such a feature of memoirs and war poetry (e.g. Blunden [1928] 1982, Sassoon [1930] 1997).

An indication of this heightened sensory experience is provided by Private Alexander Paterson who, during his time on the Western Front, developed

> . . . an expert knowledge of all the strange sounds and smells of warfare, ignorance of which may mean death . . . My hearing was attuned to every kind of explosion . . . My nostrils were quick to detect a whiff of gas or to diagnose the menace of a corpse disinterred at an interval of months. (Paterson, 1997: 239)

In such conditions, hopelessness overwhelmed many soldiers, leading them to believe every incoming shell was inscribed with a man's name (Bourke 1996: 77) – a fate perhaps deflected in popular imagination by having one's name already engraved on a talismanic bullet (and see Becker 1998: 100). The vast quantities of ordnance fired reinforced such fatalism on a grindingly regular basis. In the preliminary bombardment for the Third Battle of Ypres in July 1917, the Royal Artillery alone fired 4,283,550 rounds (Terraine 1992: 218), and, during the whole battle, the Germans discharged some 18 million shells (Werth 1997: 329). Soldiers on both sides inhabited landscapes whose terrible sights and associations yielded not only an inexhaustible supply of raw materials for metal Trench Art – but guaranteed the finished objects themselves would be deeply ambiguous.

The metallic landscapes of the Western Front were quickly etched into historical consciousness as industrialized slaughter-houses and vast tombs for 'the missing'. Although trench warfare was largely static, battlefields were metaphysically unstable places. Once tilled for crops, the fields of Picardy, Artois, and Flanders were 'drenched with hot metal' (Terraine 1996: 9), transformed from rural idylls to unrecognizable wastelands. These other-worldly landscapes were a malodorous and bizarre mixture of 'putrefaction and ammunition, the presence of the dead among the living, literally holding up trench walls from Ypres to Verdun . . .' (Winte, 1995: 68–9).

Memoirs, newspaper reports and official accounts describe this desolation with words like 'skeleton', 'gaunt' and 'broken' – imagery which associates landscape, village and human corpse. The result was '. . . a close connection, an osmosis between the death of men, of objects, of places.' (Audoin-Rouzeau 1992: 81).

Bomb-shattered churches were places of spiritual unease for Catholic French soldiers – in one cemetery, the face of a shell-shattered Christ was observed 'dripping with rain, [appearing] to reflect an infinity of suffering and sadness' (*La Saucisse* June 1917, quoted in Audoin-Rouzeau 1992: 85). The materiality of such emotions was objectified in the distinctive and often talismanic kind of metal Trench Art known as the 'bullet crucifix:' – typically made from several bullet cartridges with a crucified Christ attached (Fig. 7.7) (and see Becker, 1994: Fig. 8).

The bullet-crucifix's wider symbolic dimensions associated its miniature form variously with the pre-war roadside calvaries of France and Belgium, their wartime use as markers by soldiers (e.g. Spagnoly and Smith 1997: 11–12; and see Becker 1998: x), the 'miraculous' survival of some – such as 'Crucifix Corner' on the Somme (Middlebrook and Middlebrook 1994: 164) – and the setting up or restoration of calvaries and cruciform memorials after the war (e.g. Holt and Holt 1996: 123, Middlebrook and Middlebrook 1994: 103). The popularity of these items outlived the war to become a common souvenir for battlefield pilgrims and tourists between 1919 and 1939, with object and landscape merging with memory and loss (see later).

Figure 7.7 Bullet-crucifix, made from Mauser bullet cartridges, decorated with a commercially made Christ figure. Photo: N.J. Saunders

Figure 7.8 Women and men in a shell-filling factory. Photograph courtesy of Imperial War Museum, London

Ironies of Production

The ambiguity and irony so characteristic of soldiers' attitudes to metals and metal Trench Art was also highly gendered. Trench Art began life as ordnance in munitions factories where women were transformed from the Edwardian ideal of pacific nurturers of the race to primary producers of weapons of war and mass destruction (Ouditt 1994: 77, Woollacott 1994: 7) (Fig. 7.8). Attempts to justify this drew parallels between making bombs and making babies, referring to the womb of the shell being loaded with its deadly charge (Hall Caine quoted in Ouditt, 1994: 78–9) – a modern counterpart of attitudes in traditional metalworking societies, where the creation of metal objects is explicitly associated with the processes of pregnancy and child delivery (e.g. Rowlands and Warnier 1993: 524).

Munitions work gave women social and economic independence. At the end of 1915, 400 women worked at the Woolwich Arsenal, a figure which had increased to 27,000 by November 1918 (Ouditt 1994: 72). All together, some 400,000 women left domestic service between 1914 and 1918 mainly to work in munitions factories (Braybon 1995: 148–9). Yet, as women emerged from 'below stairs' into the well-lit well-paid world of factory work (Woollacott 1994: 4–5), men descended into the dark and dangerous world of trenches and dugouts. War *matériel* – metal Trench Art's raw material – stood at the crossroads of these two movements, liberating women but entrapping men.

Women were often ambivalent about this new work. Some felt they were doing their bit: '. . . every time you fire your gun you can remember I am helping to make the shells . . .' (Alec Tweedie quoted in Ouditt 1994: 74), while others were less sure: '. . . here I was working twelve hours a day towards the destruction of other people's loved ones . . . indirectly I was responsible for death and misery' (Peggy Hamilton, quoted in Ouditt 1994: 77).

The industrialization of armament manufacture produced greater quantities of arms than ever before, and yielded dead and maimed in industrial numbers. In this sense, we can perhaps agree with Allain Bernede (1997: 91) that 'the front . . . [was] . . . nothing but the continuation of the factory.' Women and men had a distinctive relationship with the metals of war – it objectified and scarred them both, maiming men and leaving the munitionettes with the yellow skin and orange hair characteristic of TNT poisoning (Woollacott 1994: 12).

Souvenirs, Landscape and Re-made Men

The sheer numbers of 'the missing', together with the British decision not to repatriate the dead (Winter 1995: 27), ensured a continuous flow of visitors to the old Western Front (and elsewhere) during the inter-war years. A significant proportion of these appear to have returned home with Trench Art souvenirs (and see Lloyd 1994: 131). Such objects were often the only material reminder of the dead; in the distancing process between rememberer and remembered, 'the memory of the body [was] replaced by the memory of the object' (Stewart 1994: 133).

Items were purchased for various reasons: as souvenirs of a visit, as acts of worship to the deceased's memory, and of solidarity and empathy with local people for whom their loved ones had died and whose economic hardships were everywhere apparent. As Lloyd (1994: 50) observes, 'Both Belgium and France looked to tourism to help rebuild their shattered economies.' Above all, perhaps, pilgrims were spiritually reunited with the dead through acquiring Trench Art, becoming objectified in the strange conglomerations of shells, bullets, badges and shrapnel. Forever separated from the immediacy of war, pilgrims 'authenticated' their experiences through purchasing souvenirs (Stewart 1994: 134).

In a manner akin to the sacred places of traditional societies (e.g. Carmichael et al. 1994), the Western Front became a symbolic landscape of remembrance (Winter 1995) – a place where personal and cultural identities were explored and created (Tilley 1994: 15, 26). Visiting such areas was a journey beyond the killing fields, it was a voyage through time and memory, sometimes textured by letters, war poetry, memoirs and regimental diaries. It is at this juncture that Trench Art-as-souvenir intersects with landscape, enveloping the past within the present, creating meaning by its material yet metonymic relationship to its 'natural' location (Stewart 1994: 133,136, 151; see also Saunders 1999).

If, as has been suggested (Hillman and Mazzio 1997, Stewart 1994: 125, Tilley 1994), the human body is a way of relating to and perceiving the world, then the fragmentation of corpses, artefacts, and landscape, joined together to fragment reality. This is reflected in the *Historial de la Grande Guerre in Peronne*, where splendidly baroque examples of metal Trench Art are displayed in cases adjacent to graphic instances (including a video presentation) of reconstructive surgery (see Bamji 1996), including mechanical limbs – a disturbing reminder and powerful distillation of the theoretical point that objects make people just as much as people make objects (e.g. Miller 1998: 3, Pels 1998:101).

Polished Memories, Ambiguous Displays

> The souvenirs which the pilgrims collected enabled them to carry home a tangible link with the memory, or even the spirit, of the dead. (Lloyd 1994: 185)

As an integral part of the 'house-worlds' of their owners (see Sixsmith and Sixsmith 1990: 20), metal Trench Art items played a variable and as yet largely unquantified role in the working out of personal grief and loss amongst bereaved families (and see Wenzel 1980: 1,121); in the expression of personal pride in achievement and survival amongst returning soldiers, and in their recollection of deeply formative experiences (e.g. Methley 1939). For women who had produced armaments and then lost a loved one, shells and bullets were simultaneously a manifestation of economic independence and contribution to the war effort, and – transformed into Trench Art – a constant reminder of loss. For all concerned, artefacts taken home and placed in domestic space, mediated between past and present lives, moving history into private time by juxtaposing it with a personalized present (see Stewart 1994: 138).

The object's capacity to evoke different emotions was determined by whether or not the serviceman had returned. Some items which later became objectifications of loss had originally been sent home with pride as souvenirs by a soldier who later died. The complexities of this are illustrated by the following account written by a soldier to his mother.

> By now you may have received two shell cartridges . . . They are quite safe, as they are only the empty cartridges from which the shell has been fired, and when engraved they make quite nice souvenirs . . . If the maids have time they might clean them with brass polish. I should suggest putting them on the hall mantelpiece. (LC: Letter from Col. N.B. Chaffers 19/9/15)

As a material expression of the self, metal Trench Art created a social universe of shared experiences, emotions and hopes (see Friedman, 1994: 115). For the bereaved, placing a metal letter opener or pair of polished shells on the mantelpiece, in the hallway, or on a bedside dresser was a constant reminder of the deceased. Most items, being made of brass, tarnished quickly, giving rise to a domestic routine of cleaning and polishing which probably had therapeutic effects for the bereaved. So obsessive did this behaviour sometimes become that over decades personalized inscriptions were erased almost completely (Bill Abbitt, personal communciation 1998).

Elsewhere, especially in continental Europe, 'mourning' or 'memorial' jewellery was made from recycled weaponry and metals, and had similarly therapeutic qualities (Maas and Dietrich 1994). While one can only guess at the emotions and memories unleashed or contained by seeing, cleaning, or wearing these items – sometimes for upwards of 60 years – it was nevertheless ironic that objects made for killing should be tended and regarded so lovingly.

As Whalen (1984: 37) notes for post-war Germany,

> . . . a vast amount of the war victim's energy was directed toward comprehending what death had done to them, toward enclosing the experience of violence and death in some sort of symbolic system.

Metal Trench Art was part of that system, at least in Britain, and probably also in France and Belgium (and see Saunders n.d.b).

Changing Resonances, Retailing Memories

The ideas and values which Great War Trench Art symbolized faded with the Second World War for several reasons. Many soldiers who survived the First World War were killed in the Second, remembered for the conflict which killed them not the one they survived. After 1945, the British public was much concerned with mourning a new generation of the dead and in reconstructing their towns and cities. Memories of the dead of both wars were conflated in acts of joint remembrance, further distancing many from the unique and distinctive features of the Great War.

The 1950s and early 1960s saw the low point of public interest in the Great War, and in visits to the old Western Front. Trench Art became an unfashionable anachronism, often worth more as scrap metal than objects of Great War history. However, in the mid-1960s, with the commemorations (and notably, television coverage) of the 50th anniversary of the start of the Great War, interest increased. As numbers of visitors to Great War battlefields rose, specialist tour companies appeared (Walter 1993: 63, 67–9), and there was a concomitant renewal of interest in Trench Art – as souvenirs from abroad, and as collectors' items in the burgeoning militaria trade.

Today, the personalized values of Trench Art have largely disappeared along with the original owners and makers. Gradations of 'connective memory' have been stripped away, leaving only highly relative notions of commercial value. In Britain, dealers and collectors are sometimes a repository of anonymous memories through anecdotal information which they receive from vendors as a 'value-adding' proof of authenticity. Ambiguity however remains, as dealers, collectors, and market-stall holders are often unsure how to describe or value such objects. Some regard them as a waste of a good 'clean' shell, while others see them as interesting and valuable objects from an 80-year-old war. Market valuations fluctuate wildly. Some believe Trench Art is so abundant that the market is 'flat', while others see it as increasingly rare and accruing in value. The truth probably lies somewhere in between. Today, some objects are being made as modern souvenirs, others are being faked and passed off as originals, and some have been stolen from regimental museums (e.g. Gammons 1998).

In France and Belgium – i.e. along the old Western Front – metal Trench Art is sold alongside a miscellany of defused bullets, buttons, medals, officers' whistles and pieces of uniforms – presented as a generalized category of souvenirs at the point of sale. While some items are inevitably bought by great-grandsons and great-nephews of old soldiers, the majority appear to be purchased by school-children and visitors simply as curios or souvenirs, in much the same way as caps, badges and keyrings are acquired from nearby Euro-Disney, Parc Asterix, or the archaeological theme park at Samara (SG. n.d.).

The whole 'Western Front Experience', at least in the Somme département (see Saunders n.d.a for a comparison between French and Belgian attitudes), is now part of an integrated tourist circuit as conceived and promoted by the official tourist office at Amiens (e.g. CDTS n.d.; and see AISMG n.d., HGG 1997). The social worlds and symbolic landscapes of 80 years ago which metal Trench Art objectified, are now overlain by and integrated within a reconfigured economic and political reality. Through the constant movement and transformation of metal Trench Art, the Great War continues to be recycled, materially and symbolically.

Conclusions

The broken, fragmented world which the war produced found its miniature in metal Trench Art – portable pieces of war made into whole items of peace. These artefacts can be seen as three-dimensional manifestations of Whalen's (1984: 45) point, that the grotesque nature of the war had not been denied but rather aestheticized. The strange conglomerations of shells and bullets were modernism embodied. As Hynes (1990: 195) observed, '. . . a Modernist method that before the war had seemed violent and distorting was seen to be realistic on the Western Front. Modernism had not changed, but reality had'.

Metal Trench Art can be seen as a materialization of widely held post-war attitudes which articulated notions of dislocation and fragmentation. The war was seen as a 'gap' in history (Hynes 1990: xi, 116), time itself ruptured by a conflict which shattered not only human bodies (e.g. Bourke 1996), families and relation-ships (e.g. Audoin-Rouzeau 1992: 136–7), but wider European notions of society, civilization, art and scientific progress as well (e.g. Booth 1996; Silver 1989: 1–2, 8). As Camille Mauclair wrote, 'The war has figuratively but powerfully dug a trench between yesterday's ideas and those of today . . . We have all been thrown outside ourselves . . .' (quoted in Silver 1989: 27).

Such objects embodied the confusions of war as ambiguous weapons trans-formed into ambiguous art, each object retaining visual cues to the former lives of its constituent parts. In the debate on the representation of space and reality, metal Trench Art recalls Fernand Léger's belief that Cubism was particularly appropriate to portray life in the trenches, as in his 1917 painting *The Card Party* (Silver 1989: 79, Fig. 44). Cubism's capacity to marry modern form with war-time subjects through the agency of destruction could as well be applied to metal Trench Art. The associations are suggestive – in one sense, metal Trench Art is three-dimensional Cubism, and Cubist paintings verge on two-dimensional represent-ations of metal Trench Art recyclia. 'Disguised' as artistic souvenirs, these artefacts are associated also with the wider relationship between Cubism and camouflage (see Dellouche 1994, Kahn 1984).

In *Some Reflections of a Soldier* R.H. Tawney (1916, quoted in Hynes 1990: 116) commented how on returning home from the front he felt like a visitor among strangers, separated from family by a veil unconsciously made by those who

> are afraid of what may happen to [their] souls if you expose them to the inconsistencies and contradictions, the doubts and bewilderment, which lie beneath the surface of things.

In its strange shapes and polished surfaces, metal Trench Art symbolized the poignant confusions and ambiguities of war for those who survived. It played with definitions of materiality and intent, and revealed how the concentrated intensities of industrialized war could redefine social, spiritual, and material worlds. By 1918, the pre-war European order had lost its character as a natural phenomenon (see Bourdieu 1977: 168–9), the nexus of cause and event (Eksteins 1990: 211). One consequence of this was a dizzying rush of cultural, social, economic and political upheavals and re-alignments that characterized the inter-war years. From car-boot sales to the Imperial War Museum, from French flea markets to L'Historial de la Grande Guerre, few kinds of material culture have symbolized these momentous events so well, or have endured so long, as metal Trench Art.

Acknowledgements

For guidance, encouragement and criticism I wish to thank the following: Paul Cornish (Imperial War Museum), Peter Liddle (Liddle Collection, Leeds University), Jenny Spencer-Smith and Oliver Buckley (National Army Museum), Jane Peek and Peter Aitken (Australian War Memorial), Marion Wenzel (Bosnia-Hercegovina Heritage Rescue Trust), Mike Rowlands, Danny Miller, Suzanne Kuchler, Christopher Tilley, and Barbara Bender (University College London), Jay Winter (Cambridge University), Annette Becker (Paris X Nanterre) , Ken Dunn, James Brazier, Bill Abbitt, Steve Rarity, Ralph Thompson, Angela Kelsall, Roger Lampaert (Zillebeke), John Woolsgrove (Ieper [Ypres]), Marius Kwint (Victoria and Albert Museum, London), Lt. Col. L. Deprez-Wouts (Ieper [Ypres]), and Joe and Yvonne Lyndhurst. I am especially grateful to University College London for financial support to undertake fieldwork in France and Belgium in 1998, and to the British Academy for the award of an Institutional Fellowship. I remain solely responsible for what I have made of all this generosity of spirit and matter.

References

AISMG. (n.d.), *Guide to Museums and Sites of the Great War*. Leaflet published by Members of the Association Internationale des Sites et Musees de la Guerre de 1914–1918. Peronne.

Anon. (1998), 'The Collection in Colour', *The Poppy and the Owl* 24: 97–104.

Appadurai, A. (1986), 'Introduction: Commodities and the Politics of Value', in A. Appadurai (ed.), *The Social Life of Things*, Cambridge: Cambridge University Press, 3–63.

Audoin-Rouzeau, S. (1992), *Men at War* 1914–1918: *National Sentiment and Trench journalism in France during the First World War*, Oxford: Berg.

AWM. *Australian War Memorial Military Heraldry Catalogue Worksheets/accession numbers*, Canberra: Australian War Memorial, Department of Military Heraldry.

Bamji, A. (1996), 'Facial Surgery: the Patient's experience', in H. Cecil and P.H. Liddle (eds), *Facing Armageddon: The First World War Experienced*, London: Leo Cooper, 490–519.

Becker, A. (1994), *La Guerre et La Foi: De la mort à la mémoire 1914–1930*, Paris: Armand Colin Editeur.

—— (1998), *War and Faith: Religious Imagination in France*, 1914–1930, Oxford: Berg.

Bernède, A. (1997), 'Third Ypres and the Restoration of Confidence in the Ranks of the French Army', in P.H. Liddle (ed.), *Passchendaele in Perspective*: *The Third Battle of Ypres*, London: Leo Cooper, 324–32.

Blunden, E. ([1928] 1982), *Undertones of War*, Harmondsworth: Penguin.

Boorman, D. (1988), *At the Going Down of the Sun: British First World War Memorials*, York: Sessions.

Booth, A. (1996), *Postcards from the Trenches: Negotiating the Space between Modernism and the First World War*, Oxford: Oxford University Press.

Borg, A. (1991), *War Memorials*, London: Leo Cooper.

Bourdieu, P. (1977), *Outline of a Theory of Practice*, Cambridge: Cambridge University Press.

Bourke, (1996), *Dismembering the Male: Men's Bodies, Britain and the Great War*, London: Reaktion Books.

Braybon, G. (1995), 'Women and the War', in S. Constantine, M.W. Kirby and M.B. Rose (eds), *The First World War in British History*, London: Edward Arnold, 141–67.

Carmichael, D.L., Hubert, J., Reeves, B. and Schanche, O. (eds) (1994), *Sacred Sites, Sacred Places*, London: Routledge.

CDTS. (n.d.), *Not to be Missed: 16 Places of Interest in the Somme*, Leaflet produced by the Comité Départemental du Tourisme de la Somme. Amiens.

Cerny, C. and Seriff, S. (eds) (1996), *Recycled Re-seen: Folk Art from the Global Scrap Heap*, New York: Harry N. Abrams and Museum of New Mexico.

Clout, H. (1996), *After The Ruins: Restoring the Countryside of Northern France after the Great War*, Exeter: University of Exeter Press.

Cork, R. (1994,) *A Bitter Truth: Avant-Garde Art and the Great War*, New Haven: Yale University Press.

Dellouche, D. (1994), 'Cubisme et camouflage', in J-J. Becker, J. Winter, G. Krumeich, A. Becker and S. Audoin-Rouzeau (eds), *Guerre et cultures, 1914–1918*, Paris: Armand Colin Editeur, 239–50.

Derez, M. (1997), 'A Belgian Salient for Reconstruction: People and Patrie, Landscape and Memory', in P.H. Liddle (ed.), *Passchendaele in Perspective: The Third Battle of Ypres*, London: Leo Cooper, 437–58.

Editorial. (1996), *Journal of Material Culture* 1(1): 5–14.

Eksteins, M. (1990), *The Rites of Spring: The Great War and the Birth of the Modrn Age*, New York: Anchor Books/Doubleday.

Friedman, J. (1994), *Cultural Identity and Global Process*, London: Sage Publications.

Gammons, T.J. (1998), *Letter and illustrations, dated 2/7/98*, From Regimental Headquarters The Princess of Wales's Royal Regiment (Queen's and Royal Hampshires). Canterbury.

Gough, P. (1997), '"An Epic of Mud": Artistic Impressions of Third Ypres', in P.H. Liddle (ed.), *Passchendaele in Perspective: The Third Battle of Ypres*, London: Leo Cooper, 409–21.

Harries, M. and Harries, S. (1983), *The War Artists*, London: Michael Joseph and Tate Gallery.

HGG. (1997), *Visitez L'Historial de la Grande Guerre: Premier musée internat-ional trilingue sur la Première Guerre Mondiale en Europe*, Leaflet produced by the Historial de la Grande Guerre. Péronne.

Hillman, D. and Mazzio, C. (1997), 'Introduction', in D. Hillman and C. Mazzio (eds), *The Body in Parts: Fantasies of Corporeality in Early Modern Europe*, London: Routledge, xi–xxix.

Holmes, C. ed. (1918), *The War Depicted by Distinguished British Artists*, London: The Studio.

Holt, T. and Holt, V. (1996), *Major and Mrs Holt's Battlefield Guide to the Somme*, London: Leo Cooper.

Howes, D. (1991), 'Introduction: To Summon All the Senses', in D. Howes (ed.), *The Varieties of Sensory Experience*, Toronto: University of Toronto Press, 3–21.

Hynes, S. (1990), *A War Imagined: The First World War and English Culture*, London: The Bodley Head.

IWM. *Imperial War Museum Inventory/accession numbers*. London: Imperial War Museum.

Kahn, E.L. (1984), *The Neglected Majority: 'Les Camoufleurs', Art History, and World War I*, Lanham: University Press of America.

Kelly, A. (1998), *Filming All Quiet on the Western Front: 'Brutal Cutting, Stupid Censors, Bigoted Politicos'*, London: I.B. Tauris.

King, A. (1998), *Memorials of the Great War in Britain: The Symbolism and Politics of Remembrance*, Oxford: Berg.

Kopytoff, I. (1986), 'The 'Cultural Biography of Things: Commoditization as Process', in A. Appadurai (ed.), *The Social Life of Things*, Cambridge: Cambridge University Press, 64–91.

Kwint, M. (1998), 'Images and Reflections: Pleasure and Pastime, 1914–1918', in *People's Century: Continuity Series*, London and Bicester: BBC/BCS, 241–63.

LC. *Liddle Collection/reference numbers*, The Liddle Collection, Brotherton Library, University of Leeds.

Liddle, P.H. and Richardson, M. (1997), 'Passchendaele and Material Culture: the Relics of Battle', in P.H. Liddle (ed.), *Passchendaele in Perspective: The Third Battle of Ypres*, London: Leo Cooper, 459–66.

Lloyd, D.W. (1994), 'Tourism, Pilgrimage, and the Commemoration of the Great War in Great Britain, Australia and Canada, 1919–1939', PhD Thesis, Cambridge University.

Lyndhurst, J. (1983), *Military Collectibles: An International Directory of Twentieth-Century Militaria*, Leicester: Magna Books/Salamander Books.

Maas, B. and Dietrich, G. (1994), *Lebenszeichen: Schmuck aus Notzeiten* ('Life-tokens: Jewellery for Distressing Times'). Koln: Museum fur Angewandte Kunst.

May, H.E. (1997), 'In a Highland Regiment, 1917–1918', in J.E. Lewis (Introd.) *True World War 1 Stories: Sixty Personal Narratives of the War*, London: Robinson Publishing, 199–206.

Methley, J.L.H. (1939), 'His Wallet is Stuffed with Souvenirs: A Wartime Tommy. Goes Over Some Relics', *Yorkshire Evening Post*, Friday 12 May 1939.

Middlebrook, M. and Middlebrook, M. (1994), *The Somme Battlefields: A Comprehensive Guide from Crécy to the Two World Wars*, Harmondsworth: Penguin.

Miller, D. (1985), *Artefacts as Categories*, Cambridge: Cambridge University Press.

—— (1987), *Material Culture and Mass Consumption*, Oxford: Blackwell.

—— (1998), 'Introduction', in D. Miller (ed.) *Material Cultures: Why Some Things Matter*, London: UCL Press, 3–21.

Nash, P. (1998), Letter by Paul Nash, Official War Artist, 18 November 1918, in *In Flanders Fields Museum, Cloth Hall, Market Square, Ieper: Eye Witness Accounts of the Great War: Guide to Quotations*, Ieper: Province of West Flanders.

Ouditt, S. (1994), *Fighting Forces, Writing Women: Identity and Ideology in the First World War*, London: Routledge.

Owen, W. (1993), 'A Terre', in M. Stephen (ed.) *Poems of the First World War*, London: J.M. Dent.

Paterson, A. (1997), 'Bravery in the field?', in J.E. Lewis (Introd.) *True World War One Stories: Sixty Personal Narratives of the War*, London: Robinson Publishing, 239–46.

Pels, P. (1998), 'The Spirit of Matter: On Fetish, Rarity, Fact, and Fancy', in P. Spyer (ed.), *Border Fetishisms: Material Objects in Unstable Spaces*, pp. 91–121. London: Routledge.

Richardson, M. (1996), 'Mute Witness – Material Culture in the Context of War', *The Poppy and the Owl* 19: 31–8.

Rowlands, M. and Warnier, J.P. (1993), 'The Magical Production of Iron in the Cameroon Grassfields', in T. Shaw, P. Sinclair, B. Andah and A. Okpoko (eds), *The Archaeology of Africa*, London: Routledge, 512–50.

R. W: (1915), 'Trench Trinkets: Souvenirs Soldiers make from German Shells. First Anniversary of the War Special Number; August 5th 1915', *The War Budget* IV(12): 361.

Sassoon, S. ([1930] 1997), *Memoirs of an Infantry Officer*, London: Faber and Faber.

Saunders, N.J. (1999), *Matter and Memory in the Landscapes of Conflict: The Western Front 1914–1998*, Paper prepared for symposium, 'Contested Landscapes, and Landscapes of Migration and Exile', World Archaeological Congress, Capetown, South Africa, January 1999.

—— (2000a), 'War and Material Culture', in P.H. Liddle and H; Cecil (eds), *Lightning Strikes Thrice: Personal Experiences of The World Wars*, London: Leo Cooper.

—— (2000b) 'Trench Art as Recyclia', in J. Coote (ed.), *Exhibition Catalogue on Recyclia Exhibition at the Pitt-Rivers Museum*, Oxford: Pitt-Rivers Museum.

—— (2000c), *Fighting with Style: 'Trench Art' in Sarajevo, 1999*. Paper prepared for conference, 'De Sarajevo It Sarajevo'. L'Historial de La Grande Guerre, Peronne, France, April 2000.

—— (n.d.a), *Material Memories: 'Trench Art', Transformations, and the Great War*. Mss in preparation.

—— (n.d.b), *At Home with War: Restless Objects in Unstable Places, From the Great War to Bosnia*. Mss in preparation.

Saunders, N.J. and Cornish, P. (n.d.), 'Hoarding Art, Collecting Memories: "Trench Art" from the Imperial War Museum'. Mss in preparation.

Saunders, N.J. and Wenzel, M. (n.d.), *Metalwork, Ethnicity and War: Christian – Muslim Syncretism in the 'Trench Art' Motifs and Styles of Bosnian Metalwork Tradition*. Mss in preparation.

SG. (n.d.), *Samara: The Guide*. Amiens: La Communication et La Culture du Conseil Général de la Somme.

Silver, K.E. (1989), *Esprit de Corps: Art of the Parisian Avante-garde and the First World War, 1914–25*, London: Thames and Hudson.

Simpson, K. (1996), 'Dr James Dunn and Shell-shock', in H. Cecil and P.H. Liddle (eds), *Facing Armageddon: The First World War Experienced*, London: Leo Cooper, 502–20.

Sixsmith, J. and Sixsmith, A. (1990), 'Places in Transition: the Impact of Life Events on the Experience of Home', in T. Putnam and C. Newton (eds), *Household Choices*, London: Futures Publications, pp. 20–4.

Spagnoly, T. and Smith, T. (1997), *A Walk Round Plugstreet: South Ypres Sector 1914–18*, London: Leo Cooper.

Stewart, S. (1994), *On Longing: Narratives of the Miniature, the Gigantic, the Souvenir, the Collection*, Durham, NC: Duke University Press.

Terraine, J. (1992), *White Heat: The New Warfare 1914–18*, London: Leo Cooper.

—— (1996), 'The Substance of the War', in H. Cecil and P.H. Liddle (eds), *Facing Armageddon: The First World War Experienced*, London: Leo Cooper, 3–15.

Tilley, C. (1994), *A Phenomenology of Landscape, Places, Paths and Monuments*, Oxford: Berg.

V&A/RCA (1998), *Material Memories: Design and Evocation*, Programme/synopsis booklet for Conference hosted by the Victoria and Albert Museum and the Royal College of Art at the Victoria and Albert Museum, 16–18 April 1998. London: Victoria and Albert Museum and Royal College of Art.

—— (1999), *Material Memories: Design and Evocation*, Oxford: Berg.

Walter, T. (1993), 'War Grave Pilgrimage', in I. Reader and T. Walter (eds) *Pilgrimage in Popular Culture*, Houndmills: Macmillan, 63–91.

Walters, I. (1997), 'Vietnam Zippos', *Journal of Material Culture* 2(1): 61–75.

Wenzel, M. (comp.) (1980), *Auntie Mabel's War: An Account of her Part in the Hostilities of 1914–18*. London: Allen Lane.

Werth, G. (1997), 'Flanders 1917 and the German Soldier', in P.H. Liddle (ed.), *Passchendaele in Perspective: The Third Battle of Ypres*, London: Leo Cooper, 324–32.

Whalen, R. W. (1984), *Bitter Wounds: German Victims of the Great War, 1914–1939*, Ithaca, NY: Cornell University Press.

Winter, D. (1979), *Death's Men: Soldiers of the Great War*, Harmondsworth: Penguin.

Winter, J. (1995), *Sites of Memory, Sites of Mourning: The Great War in European Cultural History*, Cambridge: Cambridge University Press.

Woollacott, A. (1994), *On Her Their Lives Depend: Munitions Workers in the Great War*, Berkeley, CA: University of California Press.

−8−

Architecture and the Domestic Sphere
Victor Buchli

Introduction

Of all the categories of material culture, architecture stands out as an artefact of great complexity, but also as the context in which most other material culture is used, placed and understood. The earliest ethnographic accounts almost always focused on architecture as one of the key diagnostic categories of material culture. Eighteenth century European accounts scrupulously noted the architectural forms of newly encountered peoples as indicators of the context of daily life and social organization and, naturally, 'progress'. The eighteenth century saw the beginnings of the concept of the primitive hut along with the noble savage as examples of idealized pristine human states, particularly in the works of the Abbé Laugier who attempted to understand emerging eighteenth century classically derived architectural forms in evolutionary terms from an idealized ethnographic 'primitive hut' (Rykwert 1989: 43–50). This category of material culture was inextricably bound up in notions of social progress as with other technological forms but more significantly: moral states of being. Portable artefacts could be divorced from their social and moral contexts, but not architecture. Built forms were almost always directly correlated with social and ethical forms. Cultures were compared to each other in terms of the social and technical complexity of their architectural forms, making architecture the single most significant artefact while describing the most fundamental unit of social organization: the family in its various guises.

The fullest expression within anthropology of the integration of built forms, social progress, and ethical states of being was probably Lewis Henry Morgan's monumental text – *Ancient Society* (Morgan 1978 [1877]) and his architectural study which followed four years later, *Houses and House life of the American Aborigines* (Morgan 1881). This latter book was supposed to have been the fifth part of *Ancient Society* which subsequently became one of the foundational texts of the newly emerging discipline of anthropology. Its fame was augmented by the immense significance it had for the development of revolutionary social thought. Marx and Engels were keen readers of his work and this Marxian relationship and its ethical dimension has almost never been lost. The architectural study, however, was separated from *Ancient Society* (Morgan 1877) through a happenstance of the pragmatics of publishing (all five together would have been too lengthy in one volume). Morgan's work on architecture, as a result, drifted into obscurity.

Victor Buchli

Morgan's two works should have been one whole culminating in an exhaustive study of the material culture of architecture as one of the corner stones of the anthropological study of human society. Ever since, however, these endeavours have tended to remain separate. In terms of the development of the discipline in the wake of figures such as Henry Maine (Carsten and Hugh-Jones 1995) the study of architecture remained marginal to social and cultural anthropology, like most other aspects of material culture studies, until the middle of the twentieth century. At this time, Edward Hall (Hall 1959) developed the theory of proxemics in the anthropological study of spatial use. But most notably, Lévi-Strauss placed architecture back into the centre of the study of human society in his work on house societies (Carsten and Hugh-Jones 1995). Architecture was brought to the foreground and understood as the social blueprint by which societies organized themselves: an empirical and durable expression of its more abstracted forms of 'social structure'. Both the 'plans' and metaphors facilitated by architecture become central to anthropological under-standings of social organization (Gudeman and Rivera 1990: 12–17; see also Oliver 1975, 1987) and latter phenomenologically influenced approaches (see Tilley this volume). Within Lévi-Strauss' work, architecture regains a centrality in anthropology not seen since Morgan. One of the significant insights of his work emphasized the importance of the house as a unifying entity that allows otherwise incompatible, antagonistic entities to cohere (Carsten and Hugh-Jones 1995) and therefore provide an exemplary context for understanding the intricacies of social dynamics structuring society. Built forms were means of living. Or, to paraphrase another of Lévi-Strauss's expressions: buildings are good to think. However Lévi-Strauss was only nominally interested in the materiality of architectural forms inasmuch as they expressed a particular 'plan' of social organization. Unlike later studies, he rarely discussed arch-itecture per se as the material artefact in its own right (Humphrey 1974, 1988, Blier 1987, Bloch 1995).

However, in the intervening period the study of architecture formed the cornerstone of most anthropological folkloric studies, especially in northern and eastern Europe representing a repository of knowledge of the titular ethnicities that were defined through the construction of nineteenth and twentieth century nation states. As much as architecture was instrumental in defining the level and complexity of social org-anization and social progress in the tradition of Morgan – nationalism required 'native' architectural forms derived from folkloric studies, archaeology and anthropology.

As such the anthropological study of architecture has focused mostly on dwellings and other vernacular forms and not on 'architecture' as the art form. The hierarchies of Western art and aesthetics which have separated Western art from so-called primitive art – the subject of anthropological interest – is similarly replicated in the archi-tectural historians' traditional preoccupation with elite and high architecture rather than the dwellings and structures of non-elites.

While the proxemic studies of Edward Hall reasserted the centrality of built forms toward the understanding of how people shaped their social relations through the study of their physical relationship to built forms, folklorists such as Henry Glassie reasserted the study of non-elite vernacular architectural forms building on the work of Hall and most innovatively the linguistic theories of Noam Chomsky (Glassie 1975).

Architecture like language was a reflection of mind, thereby reasserting the two concerns of Lévi-Strauss in a more emphatic, detailed and highly empirical mode. The linguistic turn reintroduced the significance of architecture as a language-like competence, but one that is built and used and expressed in the sensual activities of daily life as are the linguistic competences that order everyday verbal and written communication. Within British social anthropology Caroline Humphrey's highly influential piece on the Mongolian Yurt served to refocus an explicitly structuralist approach to the anthropological study of architecture as a meaningful empirical and interpretative context for understanding human societies (Humphrey 1988).

Of all the anthropological figures in the wake of Lévi-Strauss, Pierre Bourdieu's work on the study of the house amongst the Kabyle and his elaboration of his notion of habitus has been and still is one of the most significant (Bourdieu 1973, 1977). Architecture in Bourdieu's work is the literal embodiment of habitus – that unconscious set of dispositions that structure our interactions amongst ourselves and our built environment.

Within anthropological archaeology, the rise of the New Archaeology in the 1960s reintroduced the importance of the archaeological study of modern material culture especially architecture. Susan Kent's work on the ethnoarchaeological study of dwellings refocused attention on the detailed empirical study of domestic space as the most immediate material context in which to understand human societies (Kent 1990). While Ian Hodder's work in ethnoarchaelogy (Hodder 1982) along with his work in prehistory, notably *The Domestication of Europe* (Hodder 1990), placed the ethnographic and archaeological study of dwellings as part of a fully integrated study of human society with an emphasis on long-term change that would have pleased Morgan himself.

This general return of the architectural was evidenced in other related fields such as human geography which placed human habitation back into the landscape again as in the works of Amos Rappoport (Rappoport 1969). Similarly, the rise of the linguistic turn, in a more overtly Marxist and structuralist line amongst various social theorists such as Jean Baudrillard, Roland Barthes and Michel de Certeau, placed architecture and most specifically the domestic sphere of consumption as important areas of research in terms of a post-war critique of the excesses of consumer culture (Barthes 1973, Baudrillard 1996, de Certeau 1998).

Phenomenological approaches inspired by the works of Heidegger, Bachelard, Merleau-Ponty and others figure prominently in the ethnographic work of Susan Preston Blier on African architectural forms (Blier 1987). Here, built forms within this elaboration of linguistic understandings are vibrant living metaphors of social life. As Christopher Tilley relates elsewhere: 'The house is a condensed visual metaphor encapsulating essential characteristics of the cosmos' (Tilley 1999). As such, these phenomenologically inspired studies of architecture stressed the relatedness between body, dwelling and cosmos, all re-articulated through one another to assert a material and corporeal means by which to represent and live the world.

Within this climate of post-war ferment and social radicalism, feminist scholarship also refocused attention on the home and the body as the location of the ignored feminine. The legacy of Morgan through Marx and Engels forms the foundation of these feminist explorations in the spirit of Engels' famous observation : 'In the family, he is the bourgeois; the wife represents the proletariat.' (Engels 1972: 81–2). The home, its built form, and material culture was the natural place to start and address the role of women in societies. This also had quasi-forensic applications as women's experiences were often silenced as with other subaltern groups and sexualities (see Colomina 1992 and Sanders 1996). These experiences could be 'read' within prevailing linguistic understandings from the mute spaces and usages of artefacts within the domestic sphere. Thus the study of the material culture of architecture, particularly dwellings, provided an effective tool with which to diagnose social ills and inequalities and serve as a means of reforming the ethical foundations of social life within and through built forms (see Spain 1992). Henrietta Moore's work on Marakwet dwellings is a superb example of how feminist perspectives and a traditional ethno-graphic approach and sensibility towards the material forms of human life merged (Moore 1986).

This emphasis on the home as a critical site for investigation has since spawned a burgeoning interest in consumption studies based on the household and the domestic sphere, particularly in the work of Marianne Gullestad (1984) and M. Csikszentmihalyi and E. Rochberg-Halton (1981). The work of colleagues at University College London, such as Daniel Miller, has been foundational in terms of developing this emphasis into a fully fledged anthropology of consumption (Miller 1987) that emphasizes the creative cultural work of individuals in the domestic sphere rather than the cultural 'dupes' that populated previous studies of consumption and the domestic sphere (Miller 1987, 1988).

More recently the anthropological focus on dwellings has been supplanted by a broader interest in the materiality of institutional structures and cities owing a con-siderable debt to the work of Michel Foucault and his ruminations on power and built forms (Foucault 1973). A key figure is Paul Rabinow and his work on the architectural modernism of North African cities (Rabinow 1995). While Setha Low's work on cities and gated communities continues to expand the anthropologists' gaze beyond the troubled boundaries between the architecture of the home and the realm of daily life to the built environment of cities and the public sphere, focusing on the contested dynamics of contemporary society (Low 1999).

The piece presented here represents part of an effort to refocus attention away from the larger order social processes structuring Soviet society, towards the arena of the domestic sphere. Soviet Society was based on the profound reconfiguration of domestic relations, the individual, the home and daily life as part of a larger, highly contested and demiurgic process of social reform. The work is from a larger piece of research, an ethno-historical study of a socialist utopian housing complex: the Narkomfin Communal House by the Constructivist architect Moisei Ginzburg (Buchli 1999). For Soviet social reformers, the sphere of daily life ('*byt*' in Russian) and in particular the home, was the arena in which this fundamental restructuring of society was thought through and materialized, from the obviation of class antagonisms to the liberation of women. Of course this preoccupation with architecture and the home in relation to Morgan, Marx and Engels never lost its central importance to Soviet understandings (Humphrey 1988) of social life and change as an important component of 'economic-culture type' (Humphrey 1988: 17). It is not surprising that it should have been so actively problematized and seen as a key analytical context for understanding social change as well as effecting it, as we can see in the piece presented here.

Architecture in the Soviet context, was of supreme significance for the extension of state power and the generation of new forms of kinship, gender, individuation, social structure and public/private interfaces to facilitate shifting notions of socialist morality and state legitimacy as they were conceived within and lived through the Soviet home. Throughout the various attempts to intervene within the home during the Soviet period, the legacy of Morgan through Marx and Engels is always in the background. Behind all the discourses that evolved within the Soviet context the idea of Rousseau's noble savage of eighteenth century ethnographic imagination and the primitive hut of the Abbé Laugier echo throughout with the constant search for authentic forms of human interaction and ideal moral states. However, within these architectural redescriptions arose new divisions and antagonisms as illustrated in the piece here.

In short, the study of architecture enables us to think through what kinds of subjects can be imagined or not as in the Soviet example here or what is or is no longer socially viable – that is what is not 'housed' or unhomely (see Vidler 1992), within the 'non-places' mentioned here by Pinney and Bender and elsewhere by Marc Augé (Augé 1995) or through the dissipation of home and self through rituals of divestment (Marcoux 2001). Similarly, the fragmentation of political and social life and the failure of enlightenment era utopian thought has rendered the increasingly un-private, on-line work/home spaces of late capitalism – the 'last place that "the utopia of a renewal of perception" is allowed to go' (Cullens 1999: 221) – more and more problematic. Anthony Vidler has recently commented on the chaotic and destabilizing effects of contemporary spatiality and our attempts to regulate it through the built environment (Vidler 2001). If, as various phenomenologists have shown, to live is to dwell, then the increasingly fragmented and contested nature of dwelling and living poses significant challenges. In such a situation, 'homeyness' (McCracken 1989) and the utopian and moral economy of the home is ever more problematic – to not be at home in the world, in a perpetual state of unease, is the modern condition as well as its hope of social renewal, which the modernist obsession with the home has always suggested – as we know to dwell is to be. This leaves anthropologists and the material culture specialist a great deal of further work to be done in terms of our understandings of the ways built forms render material and immaterial the increasingly unstable terms of social life.

References

Augé, M. (1995), *Non-Places: an Introduction to an Anthropology of Supermodernity*, London: Verso.
Bachelard, G. (1994), *The Poetics of Space*, Boston: Beacon Press.
Barthes, R. (1973), *Mythologies*, St. Albans: Paladin.

Baudrillard, J. (1996), *The System of Objects*, London: Verso.

Bloch, M (1995) 'The Resurrection of the House Amongst the Zafimaniry of Madagascar', in J. Carsten and S. Hugh-Jones (eds), *About the House*, Cambridge: Cambridge University Press.

Blier, S.P. (1987), *The Anatomy of Architecture: Ontology and Metaphor in Batammaliba Architectural Expression*, Chicago: University of Chicago Press.

Bourdieu, P. (1973), 'The Berber House' in M. Douglas (ed), *Rule and Meanings*, Harmondsworth: Penguin

—— (1977) *Outline of a Theory of Practise*, Cambridge: Cambridge University Press.

Buchli, V. (1999), *An Archaeology of Socialism*, Oxford: Berg.

Carsten, J. and Hugh-Jones, S. (eds) (1995), *About the House: Lévi-Strauss and Beyond*, Cambridge: Cambridge University Press.

Colomina, B. (ed.) (1992), *Sexuality and Space*, New York: Princeton Architectural Press.

Csikszentmihalyi, M. and Rochberg-Halton, E. (1981), *The Meaning of Things: Domestic Symbols and the Self*, Cambridge: Cambridge University Press.

Cullens, C. (1999), 'Gimme Shelter: At Home with the Millennium', *Differences* 11(2): 204–27.

De Certeau, M. (1998), *The Practise of Everyday Life*, Minneapolis: University of Minnesota Press.

Douglas, M. (1966), *Purity and Danger: an Analysis of the Concepts of Pollution and Taboo*, London: Routledge and Kegan Paul.

Engels, F., (1972), *The Origin of the Family, Private Property and the State*, New York: Pathfinder Press.

Foucault, M. (1973), *Discipline and Punish: The Birth of the Prison*, London: Tavistock.

Glassie, H. (1975), *Folk Housing in Middle Virginia*, Knoxville: The University of Tennessee Press.

Gudeman, S. and Rivera, A. (1990), *Conversations in Colombia*, Cambridge: Cambridge University Press.

Gullestad, M. (1984) *Kitchen Table Society: A Case study of the Family Life and Friendships of Young Working-Class Mothers in Urban Norway*, Oslo: Universitetsforlaget.

Hall, E.T. (1959), *The Silent Language*. Greenwich, Conn.: Fawcett Publications.

Hodder, I. (1982), *Symbols in Action: Ethnoarchaeological Studies in Material Culture*, Cambridge: Cambridge University Press.

—— (1990), *The Domestication of Europe*, Oxford: Blackwell Publishers.

Humphrey, C. (1974), Inside a Mongolian Yurt. *New Society* 31: 45–8

Humphrey, C. (1988), 'No place like home in anthropology', *Anthropology Today* 4(1): 16–18.

Johnson, M. (1993), *Housing Culture: Traditional Architecture in an English Landscape*, London: University College London Press.

Kent, S. (1990), *Domestic Architecture and the Use of Space: an Interdisciplinary Cross Cultural Study*, Cambridge: Cambridge University Press.

Low, S. (ed.) (1999), *Theorizing the City*, New Brunswick, NJ: Rutgers University Press.

Marcoux, J-S. (2001), 'The "Casser Maison" Ritual: Constructing the Self by Emptying the Home', *Journal of Material Culture*, 6(2): 213–35.

McCracken, G. (1989), 'Homeyness', in E. Hirschman (ed.), *Interpretive Consumer Research*, Provo, UT: Association for Consumer Research.

Merleau-Ponty, M. (1962), *The Phenomenology of Perception*, London: Routledge.

Miller, D. (1987), *Material Culture and Mass Consumption*, Oxford: Blackwell.

—— (1988), 'Appropriating the State on the Council Estate', *Man* 23: 353–72.

Moore, H. (1986), *Space, Text and Gender: An Anthropological Study of the Marakwet of Kenya*, Cambridge: Cambridge University Press.

Morgan, L.H. (1978 [1877]), *Ancient Society* , Palo Alto: New York Labor News.

—— (1881) *House and House Life of the American Aborigines,* vol. IV, *Contributions to North American Ethnology*, Washington: Government Printing Office.

Oliver, P. (1987), *Dwelling: the house across the world*, Oxford: Phaidon.

—— (ed.) (1975), *Shelter, Sign and Symbol*, London: Barrie and Jenkins.

Rabinow, P. (1995), *French Modern: Norms and Forms of the Social Environment,* Chicago: University of Chicago Press.

Rappoport, A (1969), *House Form and Culture*, Englewood Hills, NJ: Prentice Hall, Inc.

Rykwert, J. (1989), *On Adam's House in Paradise: The Idea of the Primitive Hut in Architectural History*, Cambridge, MA: MIT Press.

Sanders, J. (ed.) (1996), *Stud*, New York: Princeton Architectural Press.

Spain, D. (1992), *Gendered Spaces*, Chapel Hill, N.C.: University of North Carolina Press.

Tilley, C. (1999), *Metaphor and Material Culture*, Oxford: Blackwell Publishers.

Upton, D. and Vlach, J.M. (1986), *Common Places: Readings in American Vernacular Architecture*, Athens, GA: The University of Georgia Press.

Vidler, A. (1992), *The Architectural Uncanny*, Cambridge MA: MIT Press.

—— (2000), *Warped Space*, Cambridge MA: MIT Press.

Khrushchev, Modernism and the Fight against *Petit-bourgeois* Consciousness in the Soviet Home*
Victor Buchli

Introduction

Within two years of Stalin's death, the Union of Architects released the November 1955 directive against 'ornamentalism' in the wake of the 1955 Directive of the Party and Government Against Superfluity in Project Design and Construction (Kriukovoi 1966P: 24). As for the years preceding Stalin's ascent, the years following his death resulted in a convulsion in the architectural establishment unparalleled since the victory of Stalinist Classicism over Constructivism (see Cooke 1974, Willen 1953). The Modernist principles of the 1920s were in the ascent once more. Hostilities against pre-Revolutionary domesticity that ceased under Stalin were resumed anew. The domestic realm was once more problematized, and *byt*[1] became an issue of paramount importance in post-war post-Stalinist society. The spectre of pre-revolutionary *petit-bourgeois* consciousness and its concomitant understandings of domesticity (believed by Leninist Modernists to obstruct the construction of socialism) was raised again to underline the threat of *petit-bourgeois* consciousness to the realization of full Communism in a late-socialist society. Within seven years of Stalin's death, a new cultural revolution was waged based on the principles of the first cultural revolution of the 1920s. Now, however, it was occurring unencumbered on the foundation of a fully urbanized and industrialized society. This second cultural revolution continued throughout the 1960s and decelerated in the 1970s. Reformist fervour subsided in the general malaise characterizing Soviet society throughout the course of the Brezhnev years.[2] This took hold, resulting in the final gasp of socialist reformist fervour: the Perestroika of Mikhail Gorbachev from 1985 to 1991.

The Twentieth Party Congress: 1956 and its Consequences

The campaign against ornamentalism in 1955 prefigured the start of a new cultural revolution begun in the wake of Khrushchev's 'secret speech' to the Twentieth

* Journal of Design History, no. 10, 1997, pp. 187–202

Party Congress in February 1956. Khrushchev's exposé of the horrors of the Stalinist system effectively swept away – in the minds of the Soviet public and the party – the entire preceding political and social culture. Eradicated were the events of twenty-five years and an entire generation associated with an administration's rule which made legitimate the Stalinist socialist state. This decree was no doubt a profound relief to a whole generation savaged by the vicissitudes of the Terror under Stalin's rule, particularly amongst the Intelligentsia and Nomenklatura.

This negation of the past twenty-five years was extraordinarily destabilizing and called into question the goals of the socialist project. This period of reassessment in the wake of the 'secret speech' – commonly referred to in the West as the 'Thaw' (and in recent years as the precursor to glasnost and Perestroika) – was often perceived as a period of liberalization. Some Western observers saw this as a period when the stirrings of 'free' and 'open' civil society could be felt. There was a considerable amount of open questioning at this time; but this questioning was in terms of a striving to find legitimacy for the socialist project, while recoiling from the horrific excesses of the Stalinist period. 'Freedom' in the liberal democratic sense was not what this period was about. Rather it was a search through to the roots of the socialist revolution to regain principles forsaken in the pursuit of the Stalinist state. In short it was the absolution of the Party, an absolution which could only be found in terms of the body of Bolshevik ideas which were in existence before the Stalinist ascendancy. Hence the oft-quoted slogan, 'back to Lenin'.

In terms of this reorientation, we can see how the issues of *byt*, domesticity and *petit-bourgeois* consciousness regained prominence. It assumed a position in the debates of post-war *byt* theoreticians[3] very close to the debates prevailing on the eve of the 1930 Central Committee directive which shot down all future efforts for *byt* reform under Stalin (see Buchli 1996: 68–163). In rhetorical terms, the discourse on *byt* in 1959 was virtually indistinguishable from that of 1929.

One of the most immediate consequences of the Twentieth Party Congress was the announcement towards the end of 1956 given by the Soviet of Ministers and the Central Committee of a massive building campaign. In the previous year a new administrative entity had been created, *Gosstroi* (State Construction), to oversee and co-ordinate this gargantuan effort. Housing and *byt* became paramount concerns of government and Party policy, essentially displacing in rhetorical importance earlier emphases on industrialization and the war effort.

This period should not be seen as a liberalization of attitudes towards the domestic realm. Quite the contrary. The reproblematization of *byt* was a period of intense state and Party engagement with the terms of domestic life, one that was highly rationalized and disciplined. If the Stalinist state retreated from the rationalization of the domestic 'hearth' – permitting individuals greater freedom in its expression and allowing *petit-bourgeois* consciousness to flourish – the Khrushchevist state walked straight in and began to do battle. This offensive on the domestic realm was waged on two flanks. The first was characterized by a massive building campaign. It was waged on a *tabula rasa* whereby the new post-war housing stock of the nation could be effectively inscribed with a new body of disciplining *byt* reformist principles. The second was waged on the pre-existing body of housing. It employed normative notions of good and bad taste along with household advice to regulate behaviour that would lead towards the realization of *byt* reformist socialist goals. The two flanks in turn were co-ordinated by a new ideological tool in *byt* reform, namely the emerging concept of *dizain*.

This borrowing of the English term 'design' was used to designate a co-ordinating body of normative principles regarding the nature of material culture which oversaw the rationalization of the domestic sphere in order to overcome *petit-bourgeois* consciousness. As it was explained in the popular press to Soviet readers:

> Design in its truest sense serves the harmonious self, makes it 'smart', and humanises the world in which it lives. Design teaches us to live democratically, and rationally. The importance of the appearance and phenomenon of design, of this new aspect of activity, can only be apprehended with the realisation that the world of objects has a powerful effect on people in return. That design uplifts [*vospityvaet*] not only our aesthetic sensibilities, but our ethical sensibilities as well. (Kosheleva 1976: 25)

The homophony of the terms has led many Western observers to presume that they have a similar meaning with the result that Soviet *dizain* / 'design' was seen as derivative and of little independent interest. However, *dizain* functioned very differently from its capitalist counterpart 'design', sharing only fleeting superficial similarities and consequently grossly underestimated in the Soviet context.[4]

Byt, *Dizain* and Socialist Ethics

While *byt* re-emerged as a problem almost indistinguishable in rhetorical form from its roots in the late-1920s – the field was by no means the same. The retreat from the regulation of the domestic realm under Stalin left its impact. An entire generation was raised in expectation of the joys of domesticity under terms of locally empowered action which as in other realms of social and political life characterized and assisted Stalinist state legitimacy (Buchli 1996: 198–208). In a recovering and rebuilding post-war society, these joys could not easily be denied-neither could the idea of the individual independent Stalinist agent functioning arbitrarily and locally outside objectivized Modernist norms.[5] The rhetoric of reformers often echoed the hostile tones of earlier ones from the late 1920s, often in direct contradiction with Party and government policy. An irreconcilable tension was in place between the 'on the ground', that is the 'as built' situation and the 'as legislated' directives of Party and government. The discourses of *byt* reform relied upon *byt* reformist objectives of the first cultural revolution as the only remaining legitimating socialist discourse on the domestic realm in the wake of Stalin's death and the dismantling of the cult of personality. *Byt* theory and *byt* practice were at loggerheads. The only remaining legitimating discourse on domestic practices, which in the 1920s aimed to obliterate the domestic sphere, was in direct confrontation with the expectations and aspirations towards domesticity of most Soviet citizens.

This tension was alleviated in two ways – first by understanding that Soviet society was experiencing 'high-socialism'; that it was on its penultimate phase before attaining full-blown Communism. In the 1960s that was expected to occur around 1988. Consequently Soviet society was seen to be in a transitional phase; therefore it was necessary to accommodate existing conditions. This allowed reformers to understand behaviour not corresponding to *byt* reformist ideals as temporary, regardless of how strongly individuals may have felt about their behaviour. It was simply part of an ontological phase towards full-blown Communism where domesticity would eventually become irrelevant. Secondly, normative understandings of taste began to perform an important role in regulating pre-existing patterns of domestic behaviour which flourished under Stalin and were now increasingly perceived as *petit-bourgeois*. Taste encouraged more acceptable socialist behaviour that would conform with socialist ontology. Consequently, taste and its arbiters could safely diffuse these contradictions through its regulatory functions. That is, the pursuit of domesticity could be accommodated if it followed certain rules that acknowledged this ultimate socialist ontology. Arbiters of taste could pick up those strivings for domesticity and redirect them in a manner that would eventually see them whither away as full-blown Communism was reached and the domestic 'hearth' completely obliterated. Much of the rhetoric therefore revolved around the rationalization of what *byt* theoreticians referred to as 'irrational consumer behaviour' (see Zarinskaia 1987) Thus Stalinist excess could be adequately contained by means of this disciplining regime of taste.

Needless to say, in strict terms of design, minimalism was the operative word in the arbitration of taste. This minimalism was very reminiscent of the Nordic Modernism of the West in the 1950s and 1960s – a point many Western scholars maintain, insisting that Soviet *dizain* was derivative. That copying occurred was without question but for the majority of Soviet citizens outside of Moscow these trends were certainly perceived as indigenous and thoroughly Soviet. However, if Soviet designers – both industrial and architectural as well as other arbiters of taste – looked at Modernist Western trends (they certainly were not looking at other non-Modernist trends in Western design), they were simultaneously looking at those same Modernist elements that emerged from the Cultural Revolution of the 1920s. It was no accident that theoreticians such as Khan-Magomedov began to re-evaluate the accomplishments of the pre-Stalinist generation of Soviet designers (see Khan-Magomedou 1979, 1983) In any case, the superficial appropriation of certain Western principles by Soviet designers does not diminish the unique and socially significant manner in which the appropriations were deployed within the Soviet context.

Between 1956 and 1962, a number of institutions responsible for the socialist rationalization of the domestic realm was steadily built up. The physical infrastructure within which to realize *byt* reform was established by the foundation of *Gosstroi* in 1955 and subsequent directives of the Twentieth (1956) and Twenty-First (1959) Congresses on *byt* and the housing drive. The *tabula rasa* was established upon which the first flank of the domestic front was fought. In 1962 a Soviet of Ministers of the USSR directive entitled 'Concerning quality improvement of the production of machines and consumer goods through the inculcation of methods of artistic construction' called for the rationalization of the industrial process and co-ordination of various professional and industrial concerns in order to create a rationalizing and centralized process of *dizain* over the material goods produced by the Soviet economy (Kosheleva 1976: 25). The culmination of the rationalization of the infrastructure of *byt* reform was achieved with the foundation in 1962 of the All Union Scientific Research Institute for Technical Aesthetics (VNIITE). Its mandate was to provide the rationalizing expertise of architects, industrial designers, planners, sociologists and historians for institutions such as *Gosstroi* and other sectors of the productive economy responsible for the production of consumer goods.

The second flank of the domestic front, that it dealt with pre-existing practices and the housing stock, realized *byt* reform through the arbitration of taste. Those involved revived literature on household advice and produced numerous books and pamphlets culminating in the Great Soviet Encyclopaedia's popular *Entsiklopedia Domashnevo Khoziaistva* (*Encyclopaedia of Household Economy*) first published in 1958 and continuously republished until the collapse of the Soviet Union in 1992. This literature was complemented by the journal *Dekorativnoe Iskusstvo*, and various professional journals such as *Tekhnicheskaia Estetika* by VNIITE. The concerns of *byt* reformers were given frequent coverage in popular women's journals such as *Rabotnitsa* and *Krestianka* as well as other popular media.

To shore this flank on the domestic front further, the pre-Stalinist institution of the *domkom* (housing committee) was revived with a vigour that could have only been dreamt of by earlier *byt* reformers at the time of the first Cultural Revolution. In its new form the *domkom* became the vehicle through which Party members of a particular community could actualize *byt* reformist principles at the most immediate point of contact on the domestic front, at the level of the individual household. Altogether the rationalization of *byt* on the infrastructural level and in the sphere of taste served to regulate the domestic sphere on every imaginable front. With the establishment of the VNIITE in 1962, this assault was given its final co-ordinated form which through the 1960s and early 1970s attempted to obliterate the domestic realm and vestiges of *petit-bourgeois* consciousness, towards the ontological realization of full-blown Communism.

Ethics

Complementing the emerging literature on household advice was a new emphasis on ethics, focusing on *petit-bourgeois* consciousness, commodity fetishism and irrational consumer behaviour. If mere household advice could not induce desired *byt* reforms, ethical norms were marshalled to encourage a specific relationship with the material world that ensured the moral preconditions for *byt* reform. *Petit-bourgeois* consciousness once more, as in the 1920s became a problem in the eyes of theoreticians and ethicists. In the late 1920s *petit-bourgeois* consciousness referred to the vestiges of the capitalist past hindering the development of socialism. In its reproblematization in the late 1950s and early 1960s, it served as a gloss for Stalinism as well. Thus the phenomenon of Stalin and the cult of personality was dismissed as the result of *petit-bourgeois* consciousness not having been earlier fully routed out of the collective Soviet psyche. In many respects this was an appropriate way of describing what happened. Stalinism – in order to ensure its broader legitimacy – effectively desisted from the project of the radical restructuring of the domestic realm that preoccupied a large segment of the pre-Stalinist intelligentsia. Hence the shrill rejection by *byt* reformers of the heaviness, ornamental and 'superfluous' 'pomposity' of Stalinist architecture as *petit-bourgeois*. However, the equation of the Stalinist period with a time when *petit-bourgeois* consciousness ran riot, was never explicitly stated. Certain superficial elements intimately connected with the period (such as those previously mentioned) served as the metonymic butt of criticism. It was not till much later under the glasnost of Gorbachev's Perestroika that Party theoreticians could make explicit this connection (see Starostenko 1990).

Soviet Interiors

Eliminating the stolovaia: Mono-functional versus Multi-functional Interior Plans

One of the most significant innovations of the housing boom was the reintroduction of open plans in apartment layout. There was a considerable degree of innovation in the generation of various sorts of standard plans that could accommodate the greatest degree of family variation. These attempts were very similar to earlier Constructivist ones emphasizing flexibility and other Modernist schemes from the 1920s in Western Europe.

One of the most significant aspects of these revived open plans was the reintroduction of multi-functional zone planning. Stalinist-era apartment plans were characterized by inwardly focused, mono-functional and spatially discrete rooms, reminiscent of pre-Modernist pre-revolutionary housing (Figs 8.1 and 8.2). In the 1920s such plans, categorically decried as *petit-bourgeois*, were designed to accommodate the inwardly focused, segregated and politically reactionary bourgeois nuclear family. Though such planning was not rejected in the late 1950s and early 1960s by *byt* reformers for the same reasons, the old mono-functional layout was none the less rejected. The old apportionment of space characteristic of the Stalin era and *petit-bourgeois* interiors had furniture centripetally arranged around the perimeter of the room, focusing on the family dining table or *stolovaia* at the centre. This was illuminated by the room's only light source, usually a large silken lampshade. Such configurations were summarily rejected as 'old-fashioned' as can be seen in Fig. 8.3. Here we can see how the 'traditional' arrangement above gives way to the zoned 'modern' arrangement below. The centrality in which nuclear familial relations were expressed by the Stalinist/*petit-bourgeois* layout, was now exploded into a space divided up into rationalized functional zones as this figure shows.

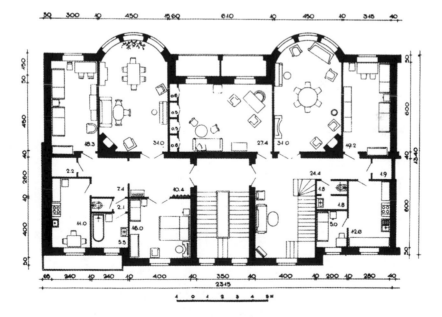

Figure 8.1 Example of a Stalin-era floor plan reminiscent of pre-revolutionary housing, 1946; State Shchusev Museum of Architecture, no 1 Zh-12497

Figure 8.2 Stalin-era interior reminiscent of pre-revolutionary interiors: Family of a Senior Metal-worker in their New Apartment, Voronezh 28 June 1948; *Tsentral'nyi Gosudarstvenyi Arkhiv Kinofotodokumentov, SSSR*, no. 0226794.

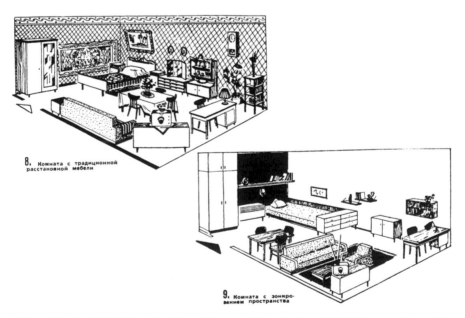

Figure 8.3 An 'old-fashioned' and *'petit-bourgeois'* interior from the Stalin era unfavourably compared with a post-Stalinist 'modern' and 'socialist' interior; V.M. Merzhanov and K.F. Sorokin, *Eto Nuzhno Novoseltsam*, Moscow Ekonomika, 1966

Most significant in this explosion of the old layout was the elimination of the dining area or *stolovaia* in its entirety. New apartment schemes accommodated dining in the kitchen, thereby relegating the family ritual of taking meals together to the 'mechanical' regions of the apartment where food was prepared and waste eliminated.[6] The *stolovaia* in effect lost its former spatial and ceremonial centrality in domestic life. If a kitchen was too small for dining, then the furniture for accommodating family meals took the form of a folding table that was stored and pulled out only when necessary (Fig. 8.4). Permanently standing furniture was placed along a wall; forming a separate functional zone with no particular spatial or symbolic emphasis (refer to lower image in Fig. 8.3).

The Children's Corner

The only area to retain spatial and symbolic significance in the new multi-functional zone plan was the corner by the window, by virtue of its direct relation

Figure 8.4 Krushchev-era kitchen with fold-away furniture; *Entsiklopedia Domashnego Khoziaistvo*, 1962

to the main source of natural light. *Byt* reformers revived an old innovation from the 1920s, the *detskii ugolok* (the children's corner) or the *ugolok diadi Lenina* (Uncle Lenin's corner) that attempted to replace the pre-revolutionary Russian Orthodox *krasnyi ugol* (icon corner). Here in the most visually prominent and well-lit corner of the room a little microcosm was created. It was a separate world in miniature that reproduced the adult space in which it was embedded (Fig. 8.5). Here the child was on constant display and under constant observation as it imbibed the Communist literary classics, surrounded by images of the nation and other instruments of edification. If patriarchal authority was enforced by reference to elders and Russian Orthodox tradition, socialist authority was enforced in reference to youth and a Communist future.

Figure 8.5 Krushchev-era 'children's corner'; *Entsiklopedia Domashnego Khoziaistvo*, 1959

Transformable Furniture: Accommodating Contradiction

One of the significant innovations in the material culture of the interior was the development of transformable furniture. This had importance in two ways for the *byt* reformist interior of the 1960s. First, it physically minimized the number of artefacts (i.e. furniture) to be found in the domestic sphere, by using one piece of furniture to serve functions that two or three separate pieces might have in the past. In many respects this was part of the general process later *byt* reformers in the late 1970s referred to as *razve shchestvleniia* (the de-artefactualization) of the domestic sphere (see Travin 1979). This was the intentional and gradual evanescence of physical objects supporting the domestic sphere that like the state would wither away with the realization of full Communism.

Second, transformable furniture served to mask functions that were considered inappropriate and secondary to the primary uses of certain rooms and furniture. For example, this ensured that a 'common room' appeared as one with its appropriate functional zones. The primary appearance and use of transformable furniture always corresponded to the correct representation of a common room as a place where members of the family gathered in the evening or received guests. The secondary and tertiary uses of transformable furniture always referred to functions that were not normally considered appropriate to the common room (a couch or bookshelf could convert into a bed, a cabinet into a child's bed and desk, a buffet into a study or a combination stereo/commode (into a dressing-table as in Fig. 8.6). Most significantly, the dining table could just simply disappear. Thus a single room apartment could give the appearance of an integrated and spare common room, such that the visitor might assume that the children's room and parent's bedroom were simply off the corridor somewhere by the toilet as is apparent in Fig. 8.7. In fact these functions that were visually factored out into imaginary other spaces could all be accommodated in one space.

Рис. 3. Комбинированный комод со встроенным туалетным столом и фонотекой.

Figure 8.6 Example of post-Stalinist-era transformable furniture: a stereo/commode by day can turn into a dressing table by night; *Entsiklopedia Domashnego Khoziastvo*, 1969

Рис. 2. Жилая комната в однокомнатной квартире
для семьи из трёх человек.

Figure 8.7 Post-Stalin-era single-room apartment; *Entsiklopedia Domashnego Khoziastvo*, 1969

It is clear from the production of such transformable furniture, and the insistence of *byt* reformers on their use, that a very particular image of the common room was anxiously desired. Why this image was sought above others is difficult to say precisely. On one hand, it allowed anyone with a one-room apartment to feel as though most of the time they were living in an apartment with at least two other rooms. In this respect one might be witnessing an accommodation to individual material aspirations. On the other hand, in the spirit of Khrushchev's peaceful competition with the capitalist West, such images demonstrated that Soviet consumers aspired to similar images of domestic life and prosperity. Images of Western conditions would filter through the popular press, certainly amongst design professionals and more specifically through large competitive international exhibitions such as those staged periodically at the VDNKh exhibition grounds. At quick glance such rooms appear spacious, comfortable and well appointed. The fact that they were designed in such a way as to accommodate three generations of a single family eating, sleeping, procreating and working all together in one room is never readily apparent from these images (Fig. 8.8).

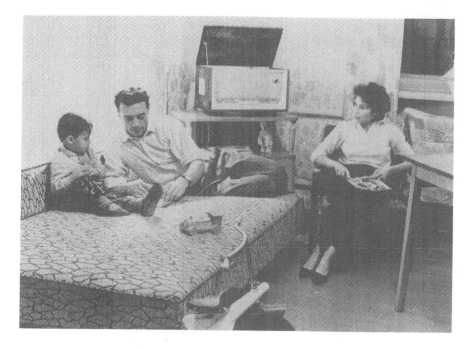

Figure 8.8 A Mechanic's Family at Home, Ust'-Kamengorsk, 1966; *Tsentral'nyi Gosudarstvenyi Arkhiv Kinofotodokumentov, SSSR*, no. 185729

Household Advice

On the second flank of the domestic front, the regulation of taste was the most important tool for the rationalization of *byt*. Taste served to regulate and modify the pre-existing housing stock and material culture of domestic artefacts. Housing and consumer goods were still in shortage during the Khrushchev years. As in the days of the First Five-Year Plan, one simply had to make do with what one had. Likewise, the pre-existing stock by virtue of its *petit-bourgeois* and now Stalinist associations contradicted the renewed ontological development of Soviet society. Just as before, one could effectively engage these vestiges of petit-bourgeois consciousness by regulating them through taste.

Much as authorities would have dearly loved for people to throw away their old furniture and buy new ones designed according to *byt* reformist principles, people only tended to buy things when they needed them at specific life intervals.[7] This meant, much to the chagrin of *byt* reformers, that whenever people moved into new apartments, they brought along with them much of the furniture they had before. As a result the new infrastructure was filled with material culture considered entirely inappropriate for its use according to *byt* reformers. Taste in this respect was marshalled to discourage the importation of vestigial material culture into these new environments. When that could not be achieved, household advice stepped in to provide guidance as to how to physically alter pre-existing furniture in conformity with *byt* reformist principles.

Exploding the Centripetal Plan

Household advice re-enforced architectural attempts to explode the centripetal plan. In terms of the manipulation of material culture, the simplest way of achieving *byt* reform in the domestic realm was a basic rearrangement of furniture. It required no material inputs on the part of consumers or industry; one could simply break the nucleus of the inwardly focused *petit-bourgeois* 'hearth' by rearranging furniture, eliminating the *stolovaia* and creating dispersed zones of use in the room.

Elimination

The next step was actually to streamline the number of items of material culture within the apartment. The slogan of 1920s *byt* reformers, *nichego lishnego* (nothing superfluous), was constantly reinvoked in the 1960s. The aim was to eliminate as many items of material culture as possible associated with the *petit-bourgeois* and Stalinist domestic realm. All but the most essential of items was to be eliminated. This applied to furniture, particularly buffet cabinets. Oft-cited artefacts to be rid of included embroidered napkins, wall rugs, figurines and various other knick – knacks. In particular *sloniki* were singled out. These rows of seven little white elephants proliferated in Stalinist interiors and served as a metonymic gloss for Stalinist *petit-bourgeois* consciousness. These elephants shown *in situ* atop a buffet cabinet in Fig. 8.9 were the focus of heated attacks by arbiters of taste in popular magazines and newspaper editorials.

In many cases, some items of furniture one simply could not do without. People needed beds; they needed places to sit. Household advisers instructed owners to physically alter individual pieces of furniture so that they conformed to *byt* reformist practices. Large beds were to be either cut down with their posts removed or feet lowered. Frames could be eliminated leaving just the mattress. Advisers advocated mounting them on a low platform. Similarly divans were to have their high backs hacked off leaving no surface to support embroidery and figurines, such as the seven little elephants. All that should remain was a low cushioned platform for sitting (Rabotnitsa 1983 no. 3: 5).

Having exploded the centripetal plan and lowered the furniture to create the desired horizontality,[8] the next step was to change the walls, floors and window treatments. These manipulations required material inputs and were additive rather

Figure 8.9 In the Apartment of Milling-machine Operator of the Proletarskii Trud Factory V.D. Fedotov, Moscow, January 1954; Central State Archive Kinofonofotodokumentov of the City of Moscow, no. O-17908 (detail of photograph)

than subtractive (although their effect was more subtractive in appearance). However, the production of paints and wallpaper designed in accord with new *byt* reformist aesthetics was sufficient such that they were both available and cheap in price. The elaborate, ubiquitous and unofficial *alfreinaia rabota* (or *trompe-l'oeil*) of Stalinist interiors was actively discouraged as can be seen in the top image of Fig. 8.10 labelled 'incorrect'. Instead, uniform and light surfaces were promoted as in the bottom image of Fig. 8.10 suitably labelled 'correct'. Appropriate wallpaper and paint products ensured that these schemes were realized.

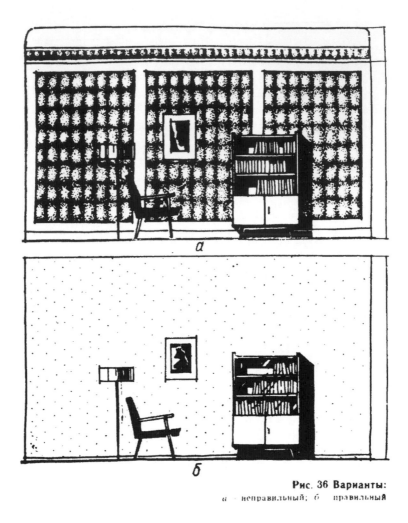

Рис. 36 Варианты:
а – неправильный; *б* правильный

Figure 8.10 'Incorrect (top) and 'correct' (bottom) wall schemes from the post-Stalin era; I.I. Serdiuk, *Kul'tura Vashoi Kvartiry*, Kiev, 1967

The manipulation of wallpapering and paint radically reconfigured interior architectural space. This achieved the most effect with the least amount of material and labour input. As the maintenance of wall surfaces was considered part of the tenant's responsibilities, the cost of this most impressive of manipulations of interior architectural space was borne directly by the tenant and not the city council or state enterprise. In addition, inspections by members of the *domkom* often addressed themselves to the need for constant interior painting and wallpapering. Guides for *domkom* party members suggested that walls be resurfaced every spring

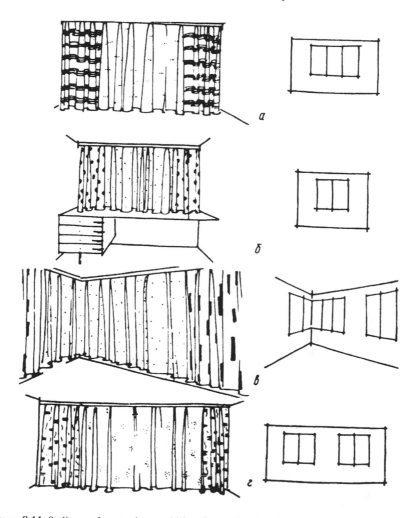

Figure 8.11 Stalin-era fenestration could be obscured and made 'modern' by the use of large expanses of drapery; I.I. Serdiuk, *Kul'tura Vashoi Kvartiry*, Kiev, 1967

ensuring the single most spatially significant alteration of domestic space by the inhabitants on a regular basis. These injunctions to redecorate were timed to coincide with Communist holidays such as May Day, and the anniversary of the October Revolution of 1917. Thus the seemingly banal activity of regular wall-papering and painting served at once to reaffirm the pursuit of domestic pleasures by a household and to give concrete expression to the construction of socialism and the legitimacy of the state.

Earlier elaborate window treatments with heavy brocaded drapes were rejected in favour of more pared-down shapes. Again, these manipulations often obscured outmoded architectural elements by covering them up with large streamlined expanses of drapery fabric as can be seen in Fig. 8.11. This at once denied the original fenestration of the architecture, whilst creating the appearance of an interior in conformity with *byt* reformist practices. In such a manner, the *byt* reformist emphasis on horizontality could be achieved by masking the verticality of architectural elements left over from the Stalinist period.

Even without the new and relatively costly inputs of *byt* reformist artefacts – such as the new furniture produced in the late 1950s and early 1960s – individuals were vigorously encouraged to realize many of the *byt* reformist principles with the minimum amount of material and labour inputs. Thus, a cosmology antithetical to a revived socialism – perceived to be inscribed within the material order – could be entirely subverted with a few strategic and ephemeral material gestures.

The Revival of the Domkom

The *domkom* had survived intact up to the late 1950s having lost much of the autonomy and radicalism which it had when first founded during the co-operative housing movements of the 1920s (see Larin 1930, Cooke 1974). As mentioned earlier, in the wake of the rapid rationalization of the domestic sphere under Khrushchev, the *domkom* received an exceptional new mandate. In 1960, a neighbourhood Leningrad Party Council announced its intention to expand its offensive on the domestic front, from the first flank of infrastructural confrontation to the second flank. One activist, Filipov, recalled this incident: 'And in those days amongst party activists the idea was born; what if one co-ordinated ones work for communist labour in production with the competition for communist norms in daily-life [*byt*], that is to say in the home?'(Filipov 1966: 151). This early initiative eventually developed into a large party movement in the larger cities of the Soviet Union. With this mandate, *domkoms* began to enact *byt* reforms directly in the domestic sphere. Quoting Filipov, 'the inhabitants of these homes collectively concern themselves with cleaning and maintenance, they participate in the improvement of the courtyards, they maintain comradely relations between themselves and actively fight against the vestiges of the old *byt*' (Filipov 1966: 154). In Moscow around the same time, the first *Dom Novogo Byta* (or House of the New Lifestyle) was organized on Bolotnikovskaia Street. The inhabitants under the stimulus of local party members organized themselves into collectives. They assumed responsibility for the maintenance of the building's landscaping. Along the balconies they cleared away all the individual household barriers and any goods privately store there. They collectively organized childcare facilities for the children of the community as well as a *stengazeta* (agitational bulletin board). In addition, Communist Party members of the *domkom* organized inspections or raids of individual apartments and recommended improvements to inhabitants.[9] All manner of collective communal services were actively promoted by the Party amongst community *domkoms* that attempted to refocus the inward looking *petit-bourgeois* household outwards on to socialist, public and communal activities. Usually these were a communal library, health clinic, children's crèche, red corner, landscaping society and other groups committed to communal activities from photo-enthusiasts to sewing circles (Aleksandrov 1968). On 7 March 1964, the Executive Committee of the Moscow City Council passed legislation permitting the organization of children's creches by *domkoms* given the approval of all parents concerned. By 1965, there were approximately 300 community-organized volunteer health clinics in operation in Moscow. According to Ionovyi and Tobinskii (refer to Aleksandrov 1968), by 1968 there were already a thousand such types of household clubs and 'red corners' all over Moscow. The Party had effectively revived an old self-supporting and self-financing institution that could efficiently service and maintain the housing stock, while simultaneously ensuring the realization of *byt* reforms. Thus the problems associated with *byt* reform before Stalinism could be finally overcome.

Conclusion

Usually, social historians have understood the Thaw as a period of liberalization after Stalinist totalitarianism. However, Stalinism could only have been possible with a limited degree of social legitimacy.[10] This legitimacy was possible in part because the state acceded a certain modicum of control in the domestic realm allowing it to exist on its own with minimal interference.[11] Far from being a period of liberalization, the Thaw witnessed a revival of a body of disciplining Modernist norms in the domestic realm. The Soviet intelligentsia and state, whose values this body of disciplinary norms most embodied, once more reasserted itself at the expense of local, individual understandings of domesticity that flourished under Stalin. Thus Modernist design principles could be effectively deployed at the most intimate level of practice: the domestic realm; effectively combating the perceived spectre, of *petit-bourgeois* consciousness. Stalinism was overturned along with its concessions to local, individually empowered agents that previously assisted Stalinist legitimacy. In its place a revived socialist project directed by a resuscitated Soviet intelligentsia ascended once more to instruct people in everyday behaviour and combat the vestiges of *petit-bourgeois* consciousness, that great bugbear of European and Soviet social reformers. These efforts had a moderating effect that began to unravel during the 'Period of Stagnation' under Brezhnev. Unable adequately to contain the contradictions of Soviet life, *byt* reform finally came undone with Perestroika and the eventual collapse of the Soviet Union. In the wake of the collapse, a new generation of elites, the so-called New Russians, have revalued the terms of the debate. Once reviled *petit-bourgeois* values are seen now as positive and socially progressive. How the terms of this discourse continue to develop remains to be seen.

Acknowledgements

I am very grateful to Susan Reid for her kind words of encouragement. I am also very indebted to Catherine Cooke for her comments, patience, and gracious help at every step. I would also like to thank the International Research and Exchanges Board for supporting research leading to this article.

Notes

1. The term *byt,* loosely translated as daily life, is an ethnographic term relating to the totality of quotidian behaviour. It refers to every aspect of daily-life, from food, clothing, domestic material culture and family life. It can also be under-stood as the English world 'lifestyle' with the additional sense of the ideological underpinnings of quotidian behaviour and material culture.
2. The malaise is referred to by Millar as an aspect of the 'little deal' under Brezhnev which resulted in a revival of acquisitive socialism reminiscent of V. Dunham's 'Big Deal' under Stalin (see Dunham 1976); see Millar, 1985.
3. '*Byt* theoreticians' here refers to a very loose, amorphous group of professional writers and activists concerned with the conditions of daily life or *byt* towards the realization of Communism in Soviet society. They were committed to the principle of democratic centralism. Thus their rhetoric could appear pluralistic in as much that debate was encouraged until which time a 'general line' could be pursued and filtered down the social hierarchy. During the Stalin period such a general line was avoided in order to ensure a modicum of popular support and integration (see Buchli 1996). *Byt* theoreticians and reformers could range from philosophers, cultural critics, literary writers, artists, household advisers and architects to party functionaries of various sorts. They were not unlike domestic reformers of the late nineteenth and early twentieth centuries who similarly ranged from socially active housewives to philosophers and architects.
4. Please see, however, Hutchings 1968 and 1976 for exceptional Western dis-cussions of Soviet design.

5. Buchli 1996: 199–208. Modernist norms here refers to the body of under-standings characteristic of the Modern Movement in architecture, design, and social and city planning in industrialized societies which attempt to regulate human behaviour according to unitary, utilitarian and rationalistic norms.
6. See Ellen Lupton 1992 for an excellent discussion of these processes in a Western context.
7. I am indebted for this observation to Gregory Prozument, Chief Designer of the former All-Union Technological Institute of Furniture Design in Moscow.
8. Refer to Hutchings (1978) on horizontality in 'Soviet design'.
9. Refer to Aleksandrov 1968: 5–7.
10. Refer to the so-called revisionist works of historians Sheila Fitzpatrick, Stephen Kotkin, and Lynn Viola.
11. Refer to Buchli 1996 and Beth Holmgren 1994 for discussions of the retreat of the Stalinist state from the domestic realm.

References

Aleksandrov, F.A. (1968), *Byt i kultura*, Moscow: Moskovskii Rabochii.

Buchli, V. (1996), 'The Narkomfin Communal House, Moscow: the material cult-ure of accommodation and resistance (1930–1991)', Ph.D. thesis, Cambridge University.

Cooke, C. (1974), 'The town of Socialism', Ph D. thesis, Cambridge University.

Dunham, V. (1976), *In Stalin's Time*, Cambridge: Cambridge University Press.

Filipov, A.P. (1966), *Kommunisticheskie nachaly v nash byt*, Leningrad: Leninizdat.

Holmgren, B. (1994), *Women's Work in Stalin's Time*, Indiana University Press.

Hudson, D. (1994), *Blueprints and Blood*, Princeton: Princeton University Press.

Hutchings, R. (1968), 'The weakening of ideological influences upon Soviet design', *Slavic Review* 27: 76–84.

—— (1976), *Soviet Science, Technology and Design*, Oxford: Oxford University Press.

—— (1978), 'Soviet design: the neglected partner of Soviet science and tech-nology', *Slavic Review*, 37(4): 567–83.

Khan-Magomedov, S.O. et al. (1979), 'Traditsii i istoki otechestvennovo dizaina', *Tekhnicheskaia estetika*, no. 21.

—— (1983), 'Nekotorye problemy razvitiia otechestvennovo dizaina'. *Tekhnich-eskaia estetika*, no. 41.

Kosheleva, I. (1976), *Rabotnitsa*, no. 3.

Kriukovoi, I.P. (1966), *Kultura zhilogo inter'ere*, Moscow: Izdatel'stvo Iskusstvo.

Larin, Iu. (1930), 'Perspektivy zhilishchnoi kooperatsii', in Iu. Larin & B. Beloussov (eds), *Za novoe zhilishche*, Moscow: Izdanie Tsentrozhilsoiuza.

Lupton, E. & Miller, J.A. (1992), The Bathroom, *The Kitchen and the Aesthetics of Waste: A Process of Elimination*, Cambridge, MA: MIT Press.

Millar, J. (1985), 'The little deal: Brezhnev's contribution to acquisitive socialism', *Slavic Review* 44(44): 694–706.

Starostenko, A.M. (1990), *Meshchianstvo kak sotsial'noe iavlenie: Genezis, sushnost', osobennosti proiavlenie preodolenia*, Moscow: Akademia Obshchestvennikh Nauk Pri TsK KPSS.

Travin, N. (1979), *Material'no-veshchnaia sreda i sotsialisticheskii obraz zhizni*, Leningrad: Nauka.

Willen, P.L. (1953), 'Soviet architecture in transformation', MA thesis, Faculty of Political Science, Columbia University.

Zarinskaia, I.Z. (1987), 'Sotsial'nye i esteticheskie aspekty formirovaniia zhiloi predmetnoi sredy', Candidate's dissertation, VNIITE.

–9–

Consumption
Daniel Miller

Introduction

One of the main arguments for the maintenance of a strong tradition of material culture studies at UCL, even during times when this was anything but fashionable, has been that they provide a means to consider and move into niches of academic enquiry that were being neglected by other branches of anthropology or indeed academia as a whole. This was conspicuously true of the study of consumption. Material Culture within the Department of Anthropology at UCL can claim to have played a major role in initiating the study of consumption in anthropology through the work of Mary Douglas, who produced one of the two foundational texts while at this department (Douglas and Isherwood 1979). This, together with the contribution of Pierre Bourdieu (1984) whose early work expresses an unusual emphasis on material culture and the practical taxonomies of everyday life, immediately became the major point of reference for subsequent work on this subject.

Both of these works stemmed in turn from the centrality of the structuralist tradition in re-establishing the role of material culture studies, with for example Lévi-Strauss emphasising the potential of objects to act as mythic forms in *The Way of the Masks* Lévi-Strauss (1982). Any attempt to re-engage with consumption meant taking seriously the larger problem of the relationship between persons and things. On reflection what is extraordinary is how, notwithstanding the rise of actual consumption after the Second World War and the sheer plethora of goods both in terms of quantity and quality that followed a rising standard of living, this whole issue was more or less ignored by the social sciences and humanities. It was not till the late 1980s that another set of books (Appadurai 1986, McCracken 1988, Miller 1987) suggested a specific commitment to the study of consumption in anthropology. Of these three books, Appadurai represented a trajectory emerging from the study of gifts and commodities in social anthropology, McCracken was concerned with the contribution of anthropology to commercial studies such as marketing, while my own book attempted to ground such studies in the core concerns of material culture.

It was not inevitable that consumption studies would include a material culture perspective. Alternative influences at the time such as Campbell (1986) were largely concerned with issues of hedonism and religion, and Bourdieu's (1984) work could easily have become absorbed within a sociological predilection for mapping social parameters such as class, with little attention to the relationship to materiality per se.

Other disciplines such as business studies continued to have their own agenda for anthropological involvement based on other aspects of consumption, e.g. Sherry (1995).

My primary concern here is to determine the particular trajectory represented by such studies at UCL within the larger development of consumption studies, and it I think it is fair to say that our approaches represent a stream that has not entirely dissolved into the larger exercise and continues to make a distinct contribution. For example consumption studies at UCL are very different from the most popular writings on this topic in the USA. Authors such as Schor (e.g. 1998) tend to work within a trajectory stemming from the original theory of consumption constructed by Veblen at the turn of the century, and to use the topic mainly to castigate the growth of materialism, individualism and hedonism associated with consumption. I would argue that has particular roots in the American tradition. It reflects concerns that can be seen in de Tocqueville's writing before the USA ever became a consumer society, and also in the Puritan concerns of its founders. This tradition continued through a long debate about the morality of spending that has been excellently documented by Horowitz (1985) and very likely reflects a context where consumption may have remained in several respects less socialized and more individualistic and competitive than is typical in European contexts. While such generalizations about distinctions between two continents are worth making, there are plenty of individual writers that transcend them. So there are books on consumption in the UK, e.g. Gabriel and Lang (1995), that look much more like the US versions, while anthropologists such as Wilk (1989), while sharing many of the US concerns with environmentalist issues, conduct similar ethnographic enquiries to those found at UCL.

Much of my own contribution to consumption studies has come through my capacity as a supervisor rather than through my own research. This can best be viewed through two collections (Miller (ed.) 1998a and Miller (ed.) 2001). Together with current work these represent seventeen PhD projects which have several character-istics in common. Firstly they are all based upon what in certain respects could be seen as quite traditional forms of anthropological enquiry. They are classic ethno-graphic studies, all demanding a minimum of a year's fieldwork, and carried out in the language of the peoples of the area. This in turn implies a claim to levels of schol-arship which are intended to deflect any assumption that the study of consumption is somehow more 'superficial' than say the study of kinship or any other branch of anthropological enquiry.

In strong contrast to the development of consumption studies elsewhere, the reasoning behind such studies suggests that what are seen as novel topics of enquiry are not necessarily best served by radically new or experimental forms of enquiry or representation. Such approaches tend to simplify and reduce the impact of studies of topics that conservatives have wanted to dismiss as outside of proper anthro-pological consideration. What is required, at least at first, are traditional forms of grounded and holistic ethnography that are devoted to examining the consequences of consumption for all aspects of people's social and cultural lives. A case in point was the decision to reject calls to study the Internet largely through chasing it on-line, and instead insist that it was traditional ethnography that demonstrated how the consumption of the Internet on-line could only be understood in its relations to conventional topics of off-line anthropological enquiry such as kinship and religion (Miller and Slater 2000).

Instead consumption studies at UCL have tried to face up to the problem that confronts the very ability to carry out these traditional forms of enquiry. In many cases a consumer society has also become a highly privatized society that does not lend itself to the traditions of participant observation or hanging out in public domains. Indeed a commitment to the ethnographic study of consumption has led in turn to a commitment to an intrusive form of fieldwork that demands a presence 'behind closed doors' that is in the midst of the private domain. The media might connect us with the global world, foods may be derived from global capitalism, but they are typically consumed within the very private domain of people's living rooms and kitchens. Such private practices may not emerge in enquiries based on eliciting publicly legitimate opinions such as those elicited by questionnaires and focus group research. The ambition represented by these projects is then to find some way for the researcher to participate within the sites where consumption as an activity is concentrated. An example of this is Tacchi's (1998) work on the consumption of radio in the home. This is a particularly private activity, especially as she concentrated on single mothers and the quite personal relationship they feel with the radio. What Tacchi thereby demonstrated is how much media research requires this kind of encounter if it is serious about understanding the consumption of media. Increasing globalization of media may be matched by increasingly private consumption.

This theme of home-based study becomes central to the second collection of student writings. In Miller (ed.) 2001 (see also Chevalier 1998) it is the home itself that becomes the focus of enquiry. Much of contemporary consumption is concerned with the home either as the object of consumption or as the setting for the arrangements and use of commodities, and the contributors to that book take a wide range of perspectives upon the relationship of homes and their possessions. These range from the topic of moving house (Marcoux 2001) and arranging the furniture (Garvey 2001b) to questioning assumptions about the tidy house in Japan (Daniels 2001) and the home as an expression of the discrepancy between aspiration and practice (Clarke 2001). I do not see how the future study of consumption can develop without this orientation to its primary site of practice. There are also obvious overlaps with the work of colleagues such as Buchli here.

The second theme that has emerged is the need to link consumption back to its relationship to production. Many of these projects along with a new interest in this topic in human geography (see Leslie and Reimer 1999) have developed what may be termed a 'commodity chain approach' which tries to use the object to trace all the links between initial production and final consumption. This has important theoretical implications, for example in my work on Coca-Cola re-printed here (Miller 1998b) I seek to challenge the important contribution of Fine and Leopold (1993) in considering these links as vertical chains of provisioning. There are many versions of how this can be done. For example Elia Petridou has worked on the links from the Greek dairy industry through wholesale and retail to consumption of dairy products in Greece. A current project by Kaori O'Conner seeks to examine the failure of the manufacturers of Lycra to fully consider the potential for this fabric amongst the baby boomer generation. Inge Daniels traced the Japanese shamoji or rice scoop from its origin in a shrine to its place in the Japanese domestic environment, to examine the relationship between commodification and spirituality. Andrew Skuse worked on both the production and consumption of an Afghan soap opera under the Taliban, and Anat Hecht has tried to consider a local history museum again from the simultaneous consideration of its production and consumption. Recently I attempted to develop an applied project based on such work, which seeks to use the Internet to involve secondary school children in a new part of the national curriculum in Geography. This had the ambition of teaching children about the people whose work is embodied in the goods they buy and more generally their responsibilities as consumers (Miller in press). In some respects this attempt at a systematic de-fetishism of the commodity completes a cycle where the enquiry into modern mass consumption at UCL began with Marx's conceptualization of the fetishism of the commodity expressed in the invisibility of its producer.

This research on re-articulating consumption and production also leads in turn to the need to re-think the connection between consumption and the wider political economy (e.g. Miller 1998c, 2001), and thereby to a long term interest within material culture at UCL with the relationship between material culture and political economy (e.g. Friedman and Rowlands 1977). Consumption studies retain a commitment to excavating their political implications. In the early stages perhaps naively it was hoped that an emphasis upon consumption would lead to a focus on the consequences of goods and services for the welfare of people. Today my own writing on political economy has tended to focus on a theory of virtualism that tries to explain why this did not in fact happen, and how various institutions have appropriated this authority of the consumer to create powerful structures such as audit and management consultancy that usurp this authority.

Other areas of concern include the process of acquisition seen most clearly in the complementarities between my own work on shopping (Miller 1998d, 2001) and that of my ethnographic partner in that enquiry, Alison Clarke, who is investigating alternative forms of provisioning and acquisition (Clarke 1998, 1999, 2000, Clarke and Miller 2002). The work on consumption has also led to a healthy interdisciplinarity that is characteristic of UCL material culture studies as a whole. We work closely with other anthropologists in the UK (e.g. Carrier 1995), scholars of consumer studies (Belk 1993), design historians (Attfield 2000), economists (Fine 1995), geographers (e.g. Miller, Jackson, Holbrook, Thrift and Rowlands 1998, Gregson 2002), psychologists (e.g. Lunt 1995) and sociologists (Miller and Slater 2000). Projects such as those of Skuse and Tacchi and the Internet research clearly overlap with media studies concerns, while other have worked on the interface with business studies, museum studies, architecture, clothing and food studies. The work has created an interesting relationship with cultural studies inasmuch as we have included much work on the consumption of contemporary cultural forms whether the Internet or cars (Garvey 2001b, Miller and Slater 2000), and worked jointly on projects (e.g. the book series *Materializing Culture* with Berg is edited with both Paul Gilroy and Michael Herzfeld). Indeed the publisher Berg has been an important player in re-articulating consumption studies and material culture more generally not only with anthropology more generally but also with disciplines such as clothing studies, gender studies and design.

The particular tradition of consumption studies at UCL is obviously not intended to constitute the whole field. It exists in complementary relation to exemplary studies elsewhere. For example in his textbook Slater (1997) integrates my own approaches within a wider sociological tradition. There is a steadily growing field of such studies in history (e.g. Campbell 1986, Davidson 1997, De Grazia with Furlough 1996, Stallybrass 1998 amongst many others), sociology (e.g. Davis (ed.) 2000, Zelizer 1997). But obviously of particular interest has been the recent growth of such studies in anthropology (e.g. Freeman 2000 in the Caribbean, Osella and Osella 1999 in South Asia, Hansen 2000 and Weiss 1996 in Africa), which reflects, as much as anything, the increasing importance of consumption within those societies traditionally studied by anthropologists. In addition anthropologists are paying more attention to this perspective on gender, ethnicity and other issues within highly developed consumer societies (e.g. Chin 2001, Layne 2000). This in turn overlaps with the growth in studies of consumption's relationship to capitalism more generally (Lien 1997, Miller 1997, Moeran 1996). Ultimately the rise of both consumption itself and the ethical, gender and environmental issues this raises (see Crocker and Linden 1998), as well as its centrality to what ought to be a return to a now neglected anthropological contribution to the study of political economy, means that the approaches discussed in this section may amount to no more than a belated acknowledgment of what is likely to increasingly becomes a central concern to all the social sciences.

References

Appadurai, A. (ed.) (1986), *The Social Life of Things: Commodities in Cultural Perspective*, Cambridge: Cambridge University Press.

Attfield, J. (2000), *Wild Things: The Material Culture of Everyday Life*, Oxford: Berg.

Auslander, L. (1996), *Taste and Power*, Berkeley: University of California Press.

Belk, R. (1993), 'Materialism and the making of the modern American Christmas', in D. Miller (ed.), *Unwrapping Christmas*, Oxford: Oxford University Press, 75–104.

Bourdieu, P. (1984), *Distinction: A Social Critique of the Judgement of Taste*, London: Routledge and Kegan Paul.

Campbell, C. (1986), *The Romantic Ethic and the Spirit of Modern Consumerism*, Oxford: Blackwell.

Carrier, J. (1995), *Gifts and Commodities*, London: Routledge.

Chevalier, S. (1998), 'From woollen carpet to grass carpet', in D. Miller (ed.), *Material Cultures. Why some Things Matter*, London : UCL Press.

Chin, E. (2001), *Purchasing Power: Black kids and American consumer culture*, Minneapolis: University of Minnesota Press

Clarke, A. (1998), '"Window Shopping at Home": Catalogues, Classifieds and New Consumer Skills', in D. Miller (ed.), *Material Cultures*, Chicago: University Of Chicago Press.

Clarke, A. (1999), *Tupperware: The Promise of Plastic in 1950s America*, Washington DC: Smithsonian Institution Press.

Clarke, A. (2000), 'Mother Swapping: the Trafficking of Second Hand Baby Wear in North London', in P. Jackson, M. Lowe, D. Miller and F. Mort (eds), *Commercial Cultures: Economies, Practices, Spaces*. Berg: Oxford.

—— (2001), 'The Aesthetics of Social Aspiration', in D. Miller (ed.), *Home Possessions*, Oxford: Berg 23–45.

Clarke, A. and Miller, D. (2002), 'Fashion and anxiety', *Fashion Theory* 6: 1–24.

Crocker, D. and Linden. T. (eds) (1998), *Ethics of Consumption: The good life, justice, and global stewardship*, Lanham: Rowman and Littlefield Publishers.

Daniels, I. (2001), 'The "untidy" House in Japan', in D. Miller (ed.), *Home Possessions*, Oxford: Berg 201–29.

Davidson, J. (1997), *Courtesans and Fishcakes*, London: Fontana Press

Davis, D. (ed.) (2000), *The Consumer Revolution in Urban China*, Berkeley: University of California Press.

De Grazia, V. with Furlough, E. (eds) (1996), *The Sex of Things*, Berkeley: University of California Press.

Douglas, M. and Isherwood, B. (1979), *The World of Good,* London: Allen Lane.

Fine, B. and Leopold, E. (1993), *The World of Consumption*, London: Routledge.

Fine, B. (1995), 'From political economy to consumption', in. D. Miller (ed.), *Acknowledging Consumption*, London: Routledge.

Freeman, C. (2000), *High Tech and High Heels in the Global Economy*, Durham, NC: Duke University Press.

Friedman, J. and Rowlands, M. (eds) (1977), *The Evolution of Social Systems*, London: Duckworth.

Gabriel T. and Lang, T. (1995), *The Unmanageable Consumer*, London: Sage.

Garvey, P. (2001a), 'Driving, drinking and daring in Norway', in D. Miller (ed.), *Car Cultures*, Oxford: Berg 133–52.

Garvey, P. (2001b), 'Organised Disorder: Moving Furniture in Norwegian Homes', in D. Miller (ed.), *Home Possessions*, Oxford: Berg 47–68.

Gregson, N. (2002), *Second Hand Worlds*, London: Routledge.

Hansen, K. T. (2000), *Salaula: The world of second hand clothing and Zambia*, Chicago: University of Chicago Press.

Horowitz D. (1985), *The Morality of Spending; attitudes towards the consumer society in America 1875–1940*, Baltimore: John Hopkins Press.

Layne, L. (2000), 'He was a real baby with baby things', *Journal of Material Culture* 5: 321–45.

Leslie, D. and Reimer, S. (1999), 'Spatializing Commodity Chains', *Progress in Human Geography* 23: 401–20.

Lévi-Strauss, C. (1982), *The Way of the Masks*, Seattle: University of Washington Press.

Lien M. (1997), *Marketing and Modernity* Oxford: Berg.

Lunt , P. (1995), 'Psychological approaches to Consumption', in D. Miller (ed.), *Acknowledging Consumption*, London: Routledge 238–63.

McCracken, G. (1988), *Culture and Consumption*, Bloomington: Indiana University Press.

Marcoux, J-S, (2001), 'The refurbishment of memory', in D. Miller (ed.), *Home Possessions*, Oxford: Berg 69–86.

Miller, D. (1987), *Material Culture and Mass Consumption*, Oxford: Blackwell.

—— (1997), *Capitalism: An Ethnographic Approach*, Oxford: Berg.

—— (ed.) (1998a), *Material Cultures*. Chicago: University of Chicago Press

—— (1998b), 'Coca-Cola. A Black sweet drink from Trinidad', in D. Miller (ed.), *Material Cultures*, Chicago: University of Chicago Press 169–87.

—— (1998c), 'A Theory of Virtualism', in Carrier and Miller (eds), *Virtualism: A new political economy*, Oxford: Berg.

—— (1998d), *A Theory of Shopping*, Cambridge: Polity/Cornell: Cornell University Press.

—— (2001), *The Dialectics of Shopping*, Chicago: University of Chicago Press.

—— (ed.) (2001), *Home Possessions*, Oxford: Berg.

—— (in press), 'Could the internet de-fetishise the commodity', *Environment and Planning D/Society and Space*.

Miller, D. Jackson, B, Holbrook, B. Thrift, N, and Rowlands, M. (1998), *Shopping, Place and Identity*, London: Routledge.

Miller, D. and Slater, D. (2000), *The Internet: An Ethnographic Approach*.

Moeran, B. (1996), *A Japanese Advertising Agency*, London: Curzon.

Osella, F. and Osella C. (1999), 'From Transience to Immanence: Consumption, Life-Cycle and Social Mobility in Kerala, South India', *Modern Asian Studies*.

Schor, J. (1998), *The Overspent American*, New York: Harper Perennial.

Sherry, J. (ed.) (1995), *Contemporary Marketing and Consumer Behaviour. An Anthropological Sourcebook*, Thousand Oaks: Sage.

Slater, D. (1997), *Consumer Culture and Modernity*, Cambridge: Polity Press.

Stallybrass P. (1998), 'Marx's Coat', in P. Spyer (ed.), *Border Fetishisms*, London: Routledge.

Tacchi, J. (1998), 'Radio Texture: Between Self and Others', in D. Miller (ed.), *Material Cultures*, Chicago: University of Chicago Press 25–45.

Weiss, B. (1996), *The Making and Unmaking of the Haya Lived World*, Durham: Duke University Press.

Wilk, R. (1989), 'Houses as consumer goods: Social processes and allocation Decisions', in H. Rutz and B. Orlove (eds), *The Social Economy of Consumption*, Lanham: University Press of America 297–322.

Zelizer, V. (1997), *The Social Meaning of Money*, Princeton, NJ: Princeton University Press.

Coca-Cola: a black sweet drink
from Trinidad*
Daniel Miller

The context for much of the current interest in material culture is a fear. It is a fear of objects supplanting people. That this is currently happening is the explicit contention of much of the debate over postmodernism which is one of the most fashionable approaches within contemporary social science. It provides the continuity between recent discussions and earlier critical debates within Marxism over issues of fetishism and reification, where objects were held to stand as congealed and unrecognized human labour. This is often an exaggerated and unsubstantiated fear, based upon the reification not of objects but of persons. It often implies and assumes a humanity that arises in some kind of pure pre-cultural state in opposition to the material world, although there is no evidence to support such a construction from either studies of the past or from comparative ethnography, where societies are usually understood as even more enmeshed within cultural media than ourselves. Rather our stance is one that takes society to be always a cultural project in which we come to be ourselves in our humanity through the medium of things.

This fear, at least in its earlier Marxist form, was not, however, a fear of material objects per se but of the commodity as vehicle for capitalist dominance, and this raises a key issue as to whether and when societies might be able to resist this particular form of object domination. Although this is a general issue, there are certain objects which have come to stand with particular clarity for this fear, and to in some sense encapsulate it. A few key commodities have come to signify the whole problematic status of commodities. Recently a new theory has been developed to consider this kind of 'meta-' status in the form of a book on the history of the swastika by Malcolm Quinn (1994). Quinn provides what he calls a theory of the meta-symbol. The swastika is unusual in that instead of standing as the icon for a specific reference it has tended to stand more generally for a meta-symbolic level

* D. Miller (ed.), *Material Cultures: Why Some Things Matter*, London: University College London Press, 1998, pp. 169–87. 1. The material for this chapter is taken in part from a larger study of business in Trinidad called *Capitalism: an ethnographic approach* (Miller 1997), which contains a wider discussion of the industry, while this chapter draws out the specific implications of Coca-Cola. A full acknowledgement of those who assisted in this project is contained in that volume. A version of this chapter has also been published in Danish in the journal *Tendens*.

that evokes the idea that there exists a higher, more mystical level of symbolization. At first after discoveries at Troy and elsewhere this was the power of symbolism in the ancient past or mystic East, but later this kind of empty but latent status allowed the Nazis to appropriate it as a generalized sign that their higher level and cultic beliefs stemmed from some deep historicity of the swastika itself. Quinn argues that the reason the swastika could achieve such importance is that whatever its particular evocation at a given time, it had come to stand above all for the general sense that there exists a symbolic quality of things above and beyond the ordinary world. This allowed people to attach a variety of mystical beliefs with particular ease to the swastika.

There may well be a parallel here to a few commodities that also occupy the position of meta-symbol. Coca-Cola is one of three or four commodities that have obtained this status. In much political, academic and conversational rhetoric the term Coca-Cola comes to stand, not just for a particular soft drink, but also for the problematic nature of commodities in general. It is a meta-commodity. On analogy with the swastika this may make it a rather dangerous symbol. It allows it to be filled with almost anything those who wish to either embody or critique a form of symbolic domination might ascribe to it. It may stand for commodities or capitalism, but equally imperialism or Americanization. Such meta-symbols are among the most difficult objects of analytical enquiry since they operate through a powerful expressive and emotive foundation such that it becomes very difficult to contradict their claimed status. So Coca-Cola is not merely material culture, it is a symbol that stands for a debate about the materiality of culture.

The title of this chapter has therefore a specific intention. It is a joke, designed to plunge us down from a level where Coke is a dangerous icon that encourages rhetoric of the type West versus Islam, or Art versus Commodity and encourages instead the slower building up of a stance towards capitalism which is informed and complex, so that any new critique has firm foundations resting on the comparative ethnography of practice within commodity worlds.

The literature on Coke that is most readily available is that which best supports the expectations raised by the meta-symbolic status of this drink. These are the many books and articles about the Coca-Cola company and its attempts to market the product. Almost all are concerned with seeing the drink as essentially the embodiment of the corporate plans of the Coca-Cola company. A good example of this is an interview published in *Public Culture* (O'Barr 1989). This is held with the head of the advertising team of the transnational agency McCann Erickson, the firm used by Coke. It focuses upon the specific question of local–global articulation. The interview traces the gradual centralization of advertising as a means to control the manner in which its image became localized regionally. In fact the advertiser notes, 'it wasn't until the late '70s that the need for advertising specifically designed for the international markets was identified.' In a sense this meant more centralization since 'the benchmark comes from the centre and anybody that needs to produce locally, for a cultural reason, a legal reason, a religious reason, or a marketing reason, has to beat it'. Many examples are given in the interview as to retakes around a core advertisement, e.g. reshot with more clothed models for Muslim countries. The interview also included discussions about the phrase 'I feel Coke' which represents a kind of Japanesed English for that market and about what attitude worked for Black Africa. Although the advertiser claims, 'I don't think Coca-Cola projects. I think that Coca-Cola reflects', the interview satisfies our desire to know what this product is by having exposed the underlying corporate strategies for global localization.

Most of the literature on the company, irrespective of whether it is enthus-iastically in favour or constructed as a diatribe against the drink, acts to affirm the assumption that the significance of the drink is best approached through know-ledge of company strategy. The literature is extensive and probably few companies and their advertising campaigns are as well documented today as Coca-Cola.

There are, however, many reasons for questioning this focus and questioning the theory of commodity power that is involved in assuming that the company controls its own effects. Indeed Coca-Cola could be argued to be a remarkably unsuitable candidate for this role as the key globalized corporation. Two reasons for this emerge through the recent comprehensive history of the company, Pendergrast's book *For God, Country and Coca-Cola* (1993). The first piece of evidence was that which Pendergrast (1993: 354–71; see also Oliver 1986) calls the greatest marketing blunder of the century. His account (and there are many others) shows clearly that the company had absolutely no idea of what the response might be to their decision to change the composition of the drink in the 1980s in response to the increasing popularity of Pepsi. The enforced restitution of Classic Coke was surely one of the most explicit examples of consumer resistance to the will of a giant corporation we have on record. After all, the company had behaved impeccably with respect to the goal of profitability. The new taste scored well in blind tests, it responded to a change in the market shown in the increasing market share of Pepsi and seemed to be a sensitive response that acknowledged the auth-ority of the consumer. Despite this, when Coke tried to change the formula, in marketing terms all hell broke loose, and the company was publicly humiliated.

The second reason why Coca-Cola is not typical of globalization is that from its inception it was based upon a system of franchising. The company developed through the strategy of agreeing with local bottling plants that they would have exclusivity for a particular region and then simply selling the concentrate to that bottler. It is only in the last few years that Coke and Pepsi have begun centralizing the bottling system and then only within the USA. There are of course obligations by the bottlers to the company. The most important are the quality control which is common to most franchising operations and the second is control over the use of the company logo. Indeed this was one of the major early sources of contention resolved in a US Supreme Court decision. In other respects the franchise system allows for a considerable degree of local flexibility, as will be shown in this chapter. Felstead (1993) documents the relationship between the specific case of Coke and more general trends in business franchising.

Coca-Cola in Trinidad

In some ways Coca-Cola is perhaps less directly associated with the States within the USA, where its presence is taken for granted, than in an island such as Trinidad, where its arrival coincided with that of an actual US presence. Coke came to Trinidad in 1939. In 1941 the British government agreed to lease certain bases in Trinidad for 99 years. As a result US troops arrived in some force. This had a profound effect upon Trinidad, no less traumatic because of core contradictions in the way in which Trinidadians perceived these events. Trinidad was already relatively affluent compared to other West Indian islands, thanks to its oil industry, but the wage levels available on the US bases were of quite a different order, leading to a mini-boom. Furthermore the US soldiers were seen as highly egalitarian and informal compared with the aloof and hierarchical British colonial authorities. In addition there was the presence of Black American soldiers and particularly a few well-remembered Black soldiers who took a sympathetic and active role in assisting the development of local Trinidadian institutions such as education.

The Americans reflected back upon themselves this benign side of the relationship in extracting the Trinidadian calypso 'Rum and Coca-Cola', which was a tremendous hit within the USA. There was, however, another side to 'Rum and Coca-Cola' that was evident within the lyrics of the calypso itself, as in the lines 'mother and daughter working for the Yankee dollar', but are found in more detail in the resentment echoed in the book with the same title written by the Trinidadian novelist Ralph de Boissiere (1984). There were many US soldiers who looked upon Trinidadians merely as a resource and were remembered as brutal and exploitative. At the time many Trinidadians felt that it was the commoditization of local sexuality and labour that was objectified through the mix of rum and Coke.

In general, however, given the ideological needs of the independence movement from Britain that followed, it was the benign side of the American presence that is remembered today. Another legacy is the drink rum and Coke which has remained ever after as the primary drink for most Trinidadians. This not only secures the market for Coke but also makes it, in this combination, an intensely local nationalist drink, whose only rival might be the beer Carib. Before reflecting on the meaning of the drink in consumption, however, we need to examine the commercial localization of Coke.

The Trinidadian Soft Drink Industry

Since Coke always works by franchise, its localization as a business comes through the selection of a local bottling plant. In Trinidad Coke is bottled by the firm of Cannings. This is one of the oldest grocery firms in the island. Established in 1912, it was described in 1922 as follows:

> His establishment became the leading place in Port of Spain for groceries, provisions, wines and spirits. It gives employment to about one hundred and ten persons. . . . It would be difficult to find anywhere a stock more representative of the world's preserved food products. . . . An example of the firm's progressive policy in all they undertake is also found in the electrical machinery equipment of their contiguous aerated water factory, where all kinds of delicious non-alcoholic beverages are made from carefully purified waters. (Macmillan 1922: 188)

The franchise for Coca-Cola was obtained in 1939 and was later expanded with the bottling of Diet Coke and Sprite. Coke by no means dominated Cannings, which continued to expand in its own right and developed with Hi-Lo what remains the largest chain of grocery stores on the island. An interesting vicarious insight into this relationship is provided by the Trinidadian novelist V.S. Naipaul (1967). The protagonist of his book *The Mimic Men* is a member of the family that own the Bella Bella bottling works which holds the Coca-Cola franchise for the semi-fictional island of Isabella. In the book a child of the owner is shown taking immense pride and personal prestige from the relationship, viewing the presence of Pepsi as a discourtesy to his family, and showing school groups around the Coke bottling works. The novel thereby provides a glimpse into the earlier localization of Coca-Cola manufacture into local circuits of status (1967: 83–6).

In 1975, Cannings was taken over, as were many of the older colonial firms, by one of the two local corporations that were becoming dominant in Trinidadian business. Unlike many developing countries which are clearly controlled by foreign transnationals, the firm of Neil & Massey along with ANSA McAl together represent the result of decades of mergers of Trinidadian firms (for details see Miller 1997: ch. 3). Both are locally owned and managed and are more than a match for the power of foreign transnationals. Indeed Neil & Massey, which took over Cannings, is clearly itself a transnational although retaining its Trinidadian base. Begun in 1958 and becoming a public company in 1975, it is now the largest firm in the Caribbean. The 1992 annual report noted assets of over TT$1.5 billion, 7,000 employees and subsidiary companies in 16 countries including the USA.[1]

Although ownership lies with Neil & Massey, the bottling section is still, in the public mind, largely identified with Cannings. This relationship is long standing and it cannot be assumed that the local company is dominated by Coke. Cannings represents reliability and quality of a kind that Coke needs from its bottlers. So while Coke would like to push Fanta as its product, and thus increase its own profits, Cannings, which makes good money on its own orange flavour, has simply refused to allow the product into Trinidad. At times it has also bottled for other companies with complementary products such as Schweppes and Canada Dry.

In effect the services offered by Coke Atlanta are largely paid for in the cost of the concentrate which is the most tax-efficient way of representing their financial relationship. In marketing, pricing, etc. the bottler has considerable autonomy to pursue its own strategies. But Cannings has little incentive to do other than follow Atlanta which will often put up 50 per cent of the costs for any particular marketing venture, and can provide materials of much higher quality than the local company could produce. As it happens McCann Erikson, which is Coca-Cola's global advertiser, is present as the sole transnational advertising agency at present operating in Trinidad.

Cannings and in turn Neil & Massey do not, however, represent localization in any simple sense, since both derive from old colonial firms. While they may therefore be seen as representing national interests as against foreign interests they may also be seen as representing white elite interests as against those of the dominant populations in the country which is split between a 40 per cent ex-African, originally mainly though by no means entirely ex-slave, and a 40 per cent ex-South Asian, almost entirely originally ex-indentured labourers (for further details on Trinidad see Miller 1994).

The importance of this identification is clarified when its competitors are brought into the picture. There are six main bottlers of soft drinks in Trinidad. Each has a specific reputation which bears on the drinks they produce. Solo is owned by an Indian family, which started in the 1930s with the wife boiling up the syrup and the husband bottling by hand. It then become one of the earlier larger firms to be outside of the control of whites and it has been either the market leader or at least a key player in the soft drink market ever since. Its main product is a flavour range called Solo and it has one franchise called RC Cola. In a sense while Solo is seen as a local "as opposed to white', the firm of S.M. Jaleel is generally regarded as more specifically 'Indian'. This is because it has the only factory in the South which is the area most dominated by the Indian population, and historically it grew from roots in a red drink that was sold mainly to Indians. Today its own flavour range is called Cole Cold but it has tried out various franchises such as for Schweppes, Seven Up and Dixi Cola.

The firm of L.J. Williams is regarded locally as the 'Chinese' bottlers. As traders and importers they are not associated with their own label but with the brand of White Rock, together with franchises for drinks including Peardrax and Ribena from Britain. The final major company is called Bottlers, though it was recently taken over by a larger firm called Amar. Bottlers have the contract for Pepsi Cola, which at times has had a significant presence in Trinidad but has been languishing with a smaller market share mainly through its inability to find ties with a reliable distributor. Its presence is mainly felt through its links with Kentucky Fried Chicken and the large amount of advertising that has continued despite its lack of market share. It also has a flavour range called USA Pop.

Marketing materials suggest that brand share would be around 35 per cent Coke, 20 per cent Solo, 10 per cent each for Cannings and Cole Cold, with the rest divided between both the other brands mentioned and various minor brands. The industry was probably worth around TT$200 million with a bottle costing 1 dollar to the consumer. The industry sees itself as an earner of foreign currency for Trinidad. This is because the only major imported element is the concentrate. Trinidad is a sugar producer; it also has a highly efficient local producer of glass bottles which manufactures for companies around the Caribbean and sometimes even for Florida. The gas is also produced as a by-product of a local factory. Soft drinks are in turn exported to a number of Caribbean islands, for example nearby Grenada is largely dominated by Trinidadian products and there is some exporting to expatriate populations in, for example, Toronto, New York and London. All the main companies involved are Trinidadian. Indeed even the local Coca-Cola representative is a Trinidadian who is responsible for the company's operations not only in Trinidad but also in five other countries including Guyana in South America. There is also a representative from Atlanta based in Puerto Rico.

When sitting in the offices of the companies concerned, the overwhelming impression is of an obsession with the competition between these firms. Indeed I argue elsewhere (Miller 1997) that it is the actions of rival companies rather than the actions of the consumers that is the key to understanding what companies choose to do. Money spent on advertising, for example, is justified in terms of one's rival's advertising budget rather than the needs of the product in the market. The competition is intense, so that the price in Trinidad is significantly lower than in Barbados where there are only three soft drink companies.

In studying the industry in detail a number of generalizations emerge that seem to command the logic of operations almost irrespective of the desires of particular companies. The first is the cola-flavours structure. Virtually all the companies involved consist of a range of flavours usually made by the companies and a franchised Coke such as Coca-Cola, Dixi Cola, RC Cola or Pepsi. The cola product has become essential to the self-respect of the company as a serious operator, but being a franchise the profits to be made are somewhat less than the flavour ranges where the concentrate is much cheaper.

The second feature seems to be the law of range expansion. Since the 1970s each year has tended to see the entry of a new flavour such as grape, pear, banana, etc., but also a more gradual increase in choice of containers, from bottles to cans to plastic, from 10 oz to 1 litre and 2 litre to 16 oz. In general the company that starts the innovation makes either a significant profit as Solo did with canning or Jaleel did with the litre bottle, or the innovation flops and a loss is incurred. Where there is success all the other companies follow behind so that the final range for the major companies is now very similar. Taking all combinations together they may produce around 60 different products.

This logic is actually counterproductive in relation to the third law, which seems to be that the key bottleneck in company success is distribution. Some of the major advances made by companies have been in finding better ways to distribute their products to the countless small retailers, known as parlours, which are often located as part of someone's house in rural areas. Jaleel, which was a pioneer here through using smaller 3 ton trucks, developed the slogan 'zero in on a Cole Cold' precisely to draw attention to its greater availability. This need to streamline delivery is, however, undermined by the increasing diversity of product, since the latter means that a greater number of crates are required to restock the needs of any particular location. This is only one example of the way in which the imperatives within the industry may act in contradiction to each other rather than as a streamlined set of strategic possibilities.

What is Trinidadian about these business operations? Well, if, as in some anthropology, local particularity is always something that derives from some prior original diversity that somehow resists the effects of recent homogenized tendencies, then the answer is 'very little'. If, however, one regards the new differentiation of global institutions such as bureaucracy and education as they are manifested in regions, as just as an authentic and important form of what might be called a posteriori diversity, then the case is not without interest. Although the various details of how business operates may not be tremendously surprising, it is actually quite distinct from any generalized models that business management with its reliance of universalized models from economics and psychology would like to promote.

In almost any area of business, for example, the conditions that control entry of new companies to the market, the situation in the Trinidadian soft drink industry would be quite different from the situation described in the business literature in, for example, the USA (e.g. Tollison et al. 1986). Partly for that reason I found that executives with business training abroad often pontificated in a manner that clearly showed little understanding of their own local conditions.

Some of the particularities, for example the degree to which the public retains knowledge about the ownership of companies or as another example the fluctuating between competition and price fixing, may have to do with the relatively small size of the market and indeed the country. Others, such as the constant link between franchised colas and local flavour groups, have to do with more fortuitous aspects of the way this industry has been developing locally. Overall this industry, as so much of Trinidadian capitalism, no more follows from general models of business and capitalism than would the particular operation of kinship in the region follow from knowing general models and theories of kinship. Profitability, like biology in kinship, may be a factor but only as manifested through politics, personal prestige, affiliation, particular historical trajectories and so forth.

To give one last example, ethnicity alone might seem an important consideration, but Solo, although Indian, is not seen as ethnic in the same way as Jaleel, which is Indian and operates in the South. Yet in terms of politics the positions are reversed. The family that owns Solo has been associated with opposition to a government that has largely been dominated by African elements and severe difficulties have been put in its way from time to time as a result, for example preventing it from introducing diet drinks; while Jaleel, being Muslim and thus a minority within a minority, has historically been associated politically with the government against the dominant Hindu group of the dominated Indian party.

To conclude – to understand the details of marketing Coke in Trinidad demands knowledge of these local, contingent and often contradictory concerns that make up the way capitalism operates locally, together with the way these affect the relationship between local imperatives and the demands emanating from the global strategists based in Atlanta. Often the net result was that Coke representatives in Trinidad were often extremely uncertain as to the best marketing strategies to pursue even when it came to choosing between entirely opposing possibilities such as emphasized its American or its Trinidadian identity.

The Consumption of Sweet Drinks

The companies produce soft drinks. The public consume sweet drinks. This semantic distinction is symptomatic of the surprising gulf between the two localized contexts. The meaning of these drinks in consumption can easily be overgeneralized and the following points may not apply well to local elites whose categories are closer to those of the manufacturers. But to understand how the mass middle-class and working-class population perceive these drinks, one needs a different starting point. For example, sweet drinks are never viewed as imported luxuries that the country or people cannot afford. On the contrary they are viewed as Trinidadian, as basic necessities and as the common person's drink.

Apart from this being evident ethnographically it is also consistent with the policy of the state. For many years the industry was under price control. This meant that prices could be raised only by government agreement and, since this severely restricted profitability, price control was thoroughly opposed by the industry. The grounds given were that this was a basic necessity for the common person and as such needed to be controlled. The reality was that to the extent that this was true (and still today consumption is around 170 bottles per person per annum), the government saw this as a politically astute and popular move. Furthermore until the IMF (International Monetary Fund) recently ended all protectionism, Coke was protected as a locally made drink through a ban on importing all foreign made soft drinks.

As a non-alcoholic beverage, sweet drinks compete largely with fruit juices and milk drinks such as peanut punch. All are available as commodities but also have home equivalents as in diluted squeezed fruit, home-made milk drinks, and sweet drinks made from water and packet crystals. Water competes inside the home, but no one but the most destitute would request water per se while ordering a meal or snack. Unlike their rivals sweet drinks are also important as mixers for alcohol.

The importance of understanding the local context of consumption as opposed to production is also evident when we turn to more specific qualities of the drinks. From the point of view of consumers, the key conceptual categories are not the flavours and colas constantly referred to by the producers. In ordinary discourse much more important are the 'black' sweet drink and the 'red' sweet drink.

The 'red' sweet drink is a traditional category and in most Trinidadian historical accounts or novels that make mention of sweet drinks it is the red drink that is referred to. The attraction of this drink to novelists is probably not only the sense of nostalgia generally but the feeling that the red drink stands in some sense for a transformation of the East Indian. While the African has become the non-marked population, the Indian has been seen as an ethnic group with its own material culture. The red sweet drink was a relatively early example of the community being objectified in relation to a commodity as opposed to a self-produced object. The red drink is the quintessential sweet drink inasmuch as it is considered by consumers to be in fact the drink highest in sugar content. The Indian population is also generally supposed to be particularly fond of sugar and sweet products and this in turn is supposed to relate to their entry into Trinidad largely as indentured labourers in the sugar cane fields. They are also supposed to have a high rate of diabetics which folk wisdom claims to be a result of their overindulgence of these preferences.

The present connotation of the red drink contains this element of nostalgia. Partly there is the reference to older red drinks such as the Jaleel's original 'red spot'. There is also the presence of the common flavour 'Kola Champagne' which is itself merely a red sweet drink.[2] Adverts that provide consumption shots will most often refer to a 'red and a roti' as the proper combination; the implication being that non-Indians also would most appropriately take a red drink with their roti when eating out, since the roti[3] has become a general 'fast food' item that appeals to all communities within Trinidad.

The centrality of the black sweet drink to Trinidadian drinking is above all summed up in the notion of a 'rum and Coke' as the core alcoholic drink for most people of the island. This is important as rum is never drunk neat or simply on the rocks but always with some mixer. Coke does not stand on this relationship alone, however. The concept of the 'black' sweet drink as something to be drunk in itself is nearly as common as the 'red' sweet drink. Coke is probably the most common drink to be conceptualized as the embodiment of the 'black' sweet drink, but any black drink will do. This is most evident at the cheapest end of the market. In a squatting community where I worked, a local product from a nearby industrial estate called 'bubble up' was the main drink. This company simply produced two drinks: a black (which it did not even bother calling a cola) and a red. People would go to the parlour and say 'gimme a black' or 'gimme a red'. At this level Coke becomes merely a high status example of the black sweet drink of my title.

This distinction between drinks relates in part to the general discourse of ethnicity that pervades Trinidadian conversation and social interaction (see Yelvington 1993). Thus an Indian talked of seeing Coke as a more white and 'white oriented people' drink. The term 'white oriented' is here a synonym for Black African Trinidadians. Many Indians assume that Africans have a much greater aptitude for simply emulating white taste and customs to become what is locally termed 'Afro-Saxons'. Africans in turn would refute this and claim that while they lay claim to white culture Indians are much more deferential to white persons.

Similarly an African informant suggested that '[a certain] Cola is poor quality stuff. It would only sell in the South, but would not sell in the North'. The implication here is that sophisticated Africans would not drink this substitute for 'the real thing', while Indians generally accept lower quality goods. In many respects there is a sense that Black culture has replaced colonial culture as mainstream while it is Indians that represent cultural difference. Thus a white executive noted that in terms of advertising spots on the radio 'we want an Indian programme, since marketing soft drinks has become very ethnic'.

The semiotic may or may not become explicit. One of the most successful local advertising campaigns in the sweet drink industry to occur during the period of fieldwork was for Canada Dry, which was marketed not as a ginger ale but as the 'tough soft drink'. The advert was produced in two versions. One had a black cowboy shooting at several bottles, as on a range, and finding that Canada Dry deflected his bullets. The other had an American Indian having his tomahawk blunted by this brand, having smashed the others. As the company told me, the idea was to cover the diversity of communities and the (as it were) 'red' Indian was adopted only after marketing tests had shown that there would be empathy and not offence from the South Asian Indian community of Trinidad.

I do not want to give the impression that there is some simple semiotic relationship between ethnicity and drinks. What has been described here is merely the dominant association of these drinks, red with Indian, black with African. This does not, however, reflect consumption. Indeed marketing research shows that if anything a higher proportion of Indians drink Colas, while Kola champagne as a red drink is more commonly drunk by Africans. Many Indians explicitly identify

with Coke and its modern image. This must be taken into account when compre-
hending the associated advertising. In many respects the 'Indian' connoted by the
red drink today is in some ways the African's more nostalgic image of how Indians
either used to be or perhaps still should be. It may well be therefore that the appeal
of the phrase 'a red and a roti' is actually more to African Trinidadians, who are
today avid consumers of roti. Meanwhile segments of the Indian population have
used foreign education and local commercial success to sometimes overtrump the
African population in their search for images of modernity and thus readily claim
an affinity with Coke.

In examining the connotations of such drinks we are not therefore exploring
some coded version of actual populations. Rather, as I argued in more detail elsewhere (Miller 1994: 257–90), both ethnic groups and commodities are better
regarded as objectifications that are used to create and explore projects of value
for the population. These often relate more to aspects or potential images of the
person than actual persons. What must be rejected is the argument of those debating about 'postmodernism' that somehow there is an authentic discourse of persons
and this is reduced through the inauthentic field of commodities. Indeed such
academics tend to pick on Coca-Cola as their favourite image of the superficial
globality that has replaced these local arguments.

Nothing could be further from the Trinidadian case. Here Coca-Cola both
as brand and in its generic form as 'black' sweet drink becomes an image that
develops as much through the local contradictions of popular culture as part of an
implicit debate about how people should be. If one grants that the red sweet drink
stands for an image of Indianness then its mythic potential (as in Lévi-Strauss
1966) emerges. This is an image of Indianness with which some Indians will
identify, some will not, and more commonly some will identify with only on certain occasions. But equally for Africans and others the identity of being Trinidadian
includes this presence of Indian as a kind of 'otherness' which at one level they
define themselves against, but at a superordinate level they incorporate as an
essential part of their Trinidadianness. The importance of the ethnicity ascribed to
drinks is that individual non-Indians cannot literally apply a piece of Indianness
to themselves to resolve this contradiction of alterity. Instead they can consume
mythic forms which in their ingestion in a sense provide for an identification with
an otherness which therefore 'completes' this aspect of the drinker's identity. To
summarize the attraction of such adverts is that Africans drinking a red sweet drink
consume what for them is a highly acceptable image of Indianness that is an
essential part of their sense of being Trinidadian.

Ethnicity is only one such dimension, where Coca-Cola as myth resolves a contradiction in value. Drinks also carry temporal connotations. Coke retains a notion
of modernness fostered by its advertising. But it has actually been a presence in
Trinidad for several generations. It has therefore become an almost nostalgic,

traditional image of being modern. For the Indians or indeed for any group where the desire not to lose a sense of tradition is complemented by a desire to feel modern, this is then an objectification of the modern that is literally very easily ingested. Solo retains its high market share precisely because it provides the opposite polarity. Although as a mass product it is less old than Coke, it is perceived as nostalgic. The Solo returnable is the old squat variety of glass bottle, and I cannot count the number of times I was regaled with the anecdote about how this was the bottle used a generation before to give babies their milk in. The desire for particular commodities are often like myth (following Lévi-Strauss 1966, rather than Roland Barthes 1973) an attempt to resolve contradictions in society and identity. This is nothing new; it is exactly the conclusion Marchand (1985) comes to with regard to his excellent study of advertising in the USA in the 1930s.

Consumption v. Production

These are just examples from the complex context of consumption which often frustrate the producers who are looking for a consistency in the population that they can commoditize. The problem was evident in a conversation with a Coke executive. He started by noting with pleasure a survey suggesting that the highly sophisticated advertising campaign currently being run by Coke was actually the most popular campaign at that time. But he then noted that what came second was a very amateur-looking ad for Det insecticide which had a particularly ugly calypsonian frightening the insects to death. His problem was in drawing conclusions from this survey that he could use for marketing.

In discussing this issue of production's articulation with consumption, it is hard to escape the constant question of 'active' or 'passive' consumers, i.e. how far consumers themselves determined the success of particular commodities. For example, I do not believe that the idea that sweet drinks are considered a basic necessity to be the result of a successful company promotion that results in people 'wasting' their money on what 'ought' to be a superfluous luxury. Nor was it an invention by government that became accepted. This assertion as to the past autonomy of the consumer derives partly from historical evidence in this case but also on the basis of what can be observed of the response to current campaigns.

Perhaps the primary target of soft drink advertising in 1988–9 was the return of the returnable bottle. Here what was at issue was precisely the industry attempting to second-guess the consumer's concern, in this case for thrift and price. The industry reasoned that its own profitability would be best served by trying to save the consumer money. But it was the executives who felt that the public 'ought' to respond to the depths of recession by favouring the returnable bottle. The problem was that despite heavy advertising by more than one company, the consumers seemed unwilling to respond to what all the agencies were loudly announcing to be the inevitable movement in the market. The campaigns were generally a failure, especially given that they were intended to collude with a trend rather than 'distort' public demand.

Companies can, of course, often respond well to complexity. They also compensate for those attributes that have become taken for granted. Thus Jaleel as a highly localized company used the most global style high-quality advertising, while Coke often tried to compensate for its given globality by emphasizing its links to Trinidad through sponsorship of a multitude of small regional events or organizations.

That, however, a gulf can exist between producer and consumer context is most evident when each rests upon different conceptualizations of the drinks. A prime example is the distinction between the sweet drink and the soft drink. Localisms such as sweet drink are not necessarily fostered by company executives who come from a different social milieu, and whose social prejudices often outweigh some abstract notion of profitability. The producers are part of an international cosmopolitanism within which high sugar content is increasingly looked upon as unhealthy. In connection with this, sweetness increasingly stands for vulgarity and in some sense older outdated traditions. The executives would wish to see themselves, by contrast, as trying to be in the vanguard of current trends.

There were, therefore, several cases of companies trying to reduce the sugar content of their drinks, and finding this resulted in complaints and loss of market share. As a result the sugar content remains in some cases extremely high. A good proportion of the cases of failure in the market that were recorded during fieldwork seemed to be of drinks with relatively high juice content and low sugar content. As one executive noted, 'we can do 10 per cent fruit as in Caribbean Cool. We are following the international trend here to higher juice, but this is not a particularly popular move within Trinidad. Maybe because it not sweet enough'.

Indeed while the executives consider the drinks in relation to the international beverages market, this may not be entirely correct for Trinidad. The term 'sweet' as opposed to 'soft' drink may have further connotations. The food category of 'sweets' and its associated category of chocolates is here a much smaller domain in the market than is found in many other areas such as Europe or the USA. Although sweets are sold in supermarkets and parlours they do not seem to be quite as ubiquitous as in many other countries and there are virtually no sweet shops per se. Given that the most important milk drink is actually chocolate milk, there is good reason to see drinks as constituting as much the local equivalent of the sweets and chocolates domain of other countries as of their soft drinks. But once again this emerges only out of the consumption context and one would have a hard time trying to convince local Trinidadian executives that they might actually be selling in liquid form that which other countries sell as solids.

It is this that justifies the point made earlier that to endeavour to investigate a commodity in its local context, there are actually two such contexts, one of production and distribution and the other of consumption. These are not the same and they may actually contradict each other to a surprising extent. There is an important general point here in that Fine and Leopold (1993) have argued with considerable force that consumption studies have suffered by failing to appreciate the importance of the link to production which may be specific to each of what they call 'systems of provision', i.e. domains such as clothing, food and utilities. What this case study shows is that while production and consumption should be linked, they may be wrong to assume that this is because each domain evolves its own local consistency as an economic process. Quite often they do not.

The reason that this is possible is that, although the formal goal of company practice is profitability, there are simply too many factors that can easily be blamed for the failures of campaigns. At best companies have marketing information such as blind tests on particular brands, and point-of-sale statistics. But even this information is little used except as *post hoc* justification for decisions that are most often based on the personal opinions and generalized 'gut' feelings of the key executives. Given peer pressure based on often irrelevant knowledge of the international beverage market, producers often manage to fail to capitalize on developments in popular culture that are available for commoditization.

Similarly consumers do not regard companies as merely functional providers. In a small island such as Trinidad consumers often have decided views, prejudices and experiences with regard to each particular company as well as their products. As such the reaction of consumers to a particular new flavour may be determined in large measure by those factors that make the consumer feel that it is or is not the 'appropriate' firm with which to associate this flavour. So there are cases of several firms trying out a new flavour until one succeeds. What the ethnography suggests is that there are often underlying reasons that one particular company's flavour tasted right.

Conclusion

This case study has attempted to localize production and consumption separately and in relation to each other. The effect has been one of relativism, using ethnography to insist upon the local contextualization of a global form. It follows that Coke consumption might be very different elsewhere. Gillespie (1995: 191–7) has analyzed the response to Coke among a group of West London youths from families with South Asian origins. The attraction to Coke is if anything greater in London than in Trinidad, but the grounds are quite different. In London, where immigration from South Asia is a much more recent experience, the focus was on the portrayal of the relative freedom enjoyed by youth in the USA as a state to be envied and emulated. There is no local contextualization to the consumption of the drink comparable to Trinidad. Furthermore the primary emphasis in London was on the advertisements rather than the drink itself.

The point of engaging in these demonstrations of relativism was declared at the beginning of this chapter to be an attempt to confront the dangers of Coke as meta-symbol. I confess I wanted to localize Coke partly because I was disenchanted with tedious anecdotes, often from academics, about Coke and global homogenization and sensed that the kind of glib academia that employed such anecdotes was possibly serving a rather more sinister end.

Anthropology seemed a useful tool in asserting the importance of a posteriori diversity in the specificity of particular capitalisms. A critical appraisal of capitalism requires something beyond the lazy term that ascribes it a purity of instrumentality in relation to profitability as goal seeking which does not usually bear closer inspection, any more than kinship can be reduced to biology. I therefore felt that prior to embarking upon a reformulated critique of capitalism it was important first to encounter capitalism as a comparative practice, not just a formal economistic logic (see Miller 1997).

But such a point could be made with many commodities. Coke is special because of its particular ability to objectify globality. This chapter has not questioned this ability, it has argued only that globality is itself a localized image, within a larger frame of spatialized identity. As has been shown within media studies (Morley 1992: 270–89), an image of the global is not thereby a universal image. The particular place of globality and its associated modernity must be determined by local setting. Indeed Caribbean peoples with their extraordinarily transnational families and connections juxtaposed with often passionate local attachments well exemplify such contradictions in the terms 'local' and 'global'. Trinidadians do not and will not choose between being American and being Trinidadian. Most reject parochial nationalism or neo-Africanized roots that threaten to diminish their sense of rights of access to global goods, such as computers or blue jeans. But they will fiercely maintain those localisms they wish to retain, not because they are hypocritical but because inconsistency is an appropriate response to contradiction.

So Coke and McDonald's are not trends, or symbolic of trends. Rather like whisky before them, they are particular images of globality that are held as a polarity against highly localized drinks such as sorrel and punch a creme, which unless you are West Indian you will probably not have heard of. Mattelart's (1991) *Advertising International* showed the commercially disastrous result when Saatchi & Saatchi believed its own hype about everything becoming global. The company nearly collapsed. In an English pub we would also find an extension of the range of potential spatial identifications from international or European lagers to ever more parochial 'real ales'. No doubt many aspiring anthropologists would choose one of the new 'designer' beers which typically derive from some exotic location, such as Mexico or Malaysia where anthropologists have a distinct advantage in boring other people with claimed knowledge of the original homeland of the beer in question. The point this brings home is that semiotics without structuralism was never much use, as Coca-Cola found when it tried to change its formula and was humiliated by the consumer.

Finally we can turn back to the meta-context of this chapter, which asserts the scholarship of such contextualization against the common academic use of Coca-Cola as glib generality. There are many grounds for favouring theoretical and comparative approaches in anthropology against a tradition of ethnography as mere parochialism, which is hard to justify in and of itself. In this case, however, there are specific grounds for an ethnography of localism because here localism makes a particular point. Vanguard academics seem to view Coca-Cola as totalized, themselves as contextualized. This chapter shows, by example, how ordinary vulgar mass consumption is proficient in sublating the general form back into specificity. By contrast, in much of the academic discussion of post-modernism as also in political rhetoric we find the totalizing of Coca-Cola as a meta-symbol. This discourse when detached from its historical and localized context comes to stand for the kind of anti-enlightenment irrationalism and aestheticism that was once the main instrument used by fascism against the rationalist tradition of the enlightenment. The point is not dissimilar to Habermas's (1987) argument in *The Philosophical Discourse of Modernity* against these same trends in modern academic thought.

The concept of meta-symbol certainly fits postmodern assertions about nothing referring to anything in particular any more. I do not accept at all that this is true for commodities such as Coke when investigated in production or consumption contexts, although this is what the academic asserts. But it might just be true of the way academics themselves use the image of Coke. First accepting its globality in a simple sense, Coke then becomes a general Capitalism, Imperialism, Americanization, etc. Then in discussions of postmodernism it may emerge as a kind of generalized symbol standing for the existence in commodities of a level of irrationalized meta-symbolic life (Featherstone 1991 provides a summary of these kinds of academic discussions). This parallels what Quinn (1994) sees happening to the swastika:

> Returning to the definition of the swastika as the 'sign of non-signability', we can see that here the image comes to represent the symbolic realm or the symbolic process per se, as a meta-symbol or 'symbol of symbolism', a status that Aryanism reflected by naming the image as the 'symbol of symbols' set apart from all others and representable only by itself (Quinn 1994: 57).

Both the Coca-Cola corporation but even more a trend in academics and politics would wish to push Coca-Cola onto this kind of plateau. In collusion with the drinkers that consume it and often the local companies that bottle it, this chapter is intended to form part of a counter movement that would push Coca-Cola downwards back into the muddy dispersed regions of black sweet drinks.

Notes

1. At the time the exchange rate was approximately TT$4.25 to US$l.
2. This particular marketing ploy of calling fizzy drinks champagne is found in Britain in the nineteenth century, when drinks were made from a syrup called twaddle. This may well be the origin of the expression 'a load of twaddle', a fact entirely unrelated to the rest of this chapter!
3. Roti is the traditional Indian unleavened bread, sold in Trinidad wrapped around some other food to make a meal.

References

Barthes, R. (1973), *Mythologies*, London: Paladin.
de Boissiere, R. (1984), *Rum and Coca-Cola*, London: Alison & Busby.
Featherstone, M. (1991), *Consumer Culture and Postmodernism*, London: Sage.
Felstead, A. (1993), *The Corporate Paradox*, London: Routledge.
Fine, B. and Leopold, E. (1993), *The World of Consumption*, London: Routledge.
Gillespie, M. (1995), *Television, Ethnicity and Cultural Change*, London: Routledge.
Habermas, J. (1987), *The Philosophical Discourse of Modernity*, Cambridge, MA: MIT Press.
Lévi-Strauss, C. (1966), *The Savage Mind*, London: Weidenfeld & Nicolson.
Macmillan, A. (1922), *The Red Book of the West Indies*, London: Collingridge.
Marchand, R. (1985), *Advertising the American Dream*, Berkeley: University of California Press.
Mattelart, A. (1991), *Advertising International*, London: Routledge.
Miller, D. (1994), *Modernity: An Ethnographic Approach*, Oxford: Berg.
Miller, D. (1997), *Capitalism: An Ethnographic Approach*, Oxford: Berg.
Morley, D. (1992), *Television Audiences and Cultural Studies*, London: Routledge.
Naipaul, V.S. (1967), *The Mimic Men*, Harmondsworth: Penguin.
O'Barr, W. (1989), 'The Airbrushing of Culture', *Public Culture* 2(1): 1–19.

Oliver, T. (1986), *The Real Coke: The Real Story*, London: Elm Tree.

Pendergrast, M. (1993), *For God, country and Coca-Cola*. London: Weidenfeld & Nicolson.

Quinn, M. (1994), *The swastika*. London: Routledge.

Tollison, R., Kaplan, D. Higgins, R. (eds) (1986), *Competition and Concentration: the Economics of the Carbonated Soft Drink Industry*, Lexington, MA: Lexington Books.

Yelvington, K. (ed.) 1993. *Trinidad Ethnicity*, London: Macmillan.

Index

Abelam, 73
academic traditions, 1–2, 5, 7, 116
access, to Stonehenge, 155, 162–7
acquisition, 240
advertising, 246–7, 255–8
advice, household, 227–8
aestheticization, war, 199–200
aesthetics, 58–9, 75–7, 83
agency, 60
'aha (sacred cord), 67
alienability, 15–18
American Anthropological Association, 119
amnesia, cultural *see* memory
ancestor houses, 35
ancestors, 71–5, 100–1
Ancient Society (1877), 3–4, 207
androgyny
    canoes, 40–1, 49
    slit drums, 50
Anglo-Saxons, 144
anthropologists, 130–1
anthropology, 115–16
    academic traditions, 1–2, 5
    and architecture, 207–11
    applied, 138
    consumption studies, 241
    of art, 57–60
    social, 7, 9–10, 12
    see also ethnography; ethnology
antiquarians, 152, 171n15
Antrobus, Sir Edmund, 154, 171n15
Apache, 128–9
apartments, layout, 220–1
appropriation, 145
    of landscape, 142, 146, 153
archaeologists, 122, 160–1
archaeology, 8, 10–11, 116, 209
    academic traditions, 1–2
    Great War, 178
architecture, 207–11
    Muslim, 123

art, 57–60, 65
Art and Agency (1988), 57–9
artefacts
    super-category, 6–8
    visual representation, 14
    see also collections
artefactuality, 18
artworks, 60, 63–4, 66
Ascherson, N., 171n14
Aubrey, John, 152
Austen, Jane, 135–6
Australian Aborigines, 130
Austronesian speaking cultures, 65, 78
authenticity, 109–10
authorship, 107
Avebury, 145, 147, 153, 169n1
Ayodhya mosque, 121
Azad, Chadra Shekhar, 95–6

Babadzan, A., 67
barkcloth technology, 70–1
Barthes, Roland, 94
Bataille, G., 12–13
battlefields, 193–4
    pilgrimage and tourism, 177, 190, 198–9
Battle of the Beanfield (1985), 155, 170n8
Bender, Barbara, 1–2, 135–8, 141–72
Benjamin, Walter, 83
bereavement, 197–8
Berger, John, 82
Berg Publishers, 240
Bernede, Allain, 196
Bharatiya Janata Party (BJP), 121–2
bias, in material culture studies, 11–12
bilum (netbags), 41, 75–6
binding, 63–79
biography
    and trench art, 192–9
    of objects, 176
birds, 33–6, 43–4
Blier, Susan Preston, 209

Index

wedding photography, 87–9
Weldon, Fay, 141
Weltbild see distance
Western Front, 185, 189–200 *passim*
windows, 228, 231–2
women, 197
   and dwellings, 210

munitions work, 195–6
wood
   canoes, 31–2, 34
   carving, 69
wrapping, 68–9

youths, of South Asian origin, 259–60